国家级一流本科专业建设点教材
浙江省普通本科高校"十四五"重点立项建设教材
浙江省省级线上一流课程配套教材
Unity官方中文课堂同名配套教材

游戏开发基础

基于Unity游戏引擎

张帆 主编
谢昊 吉娜烨 副主编

电子工业出版社
Publishing House of Electronics Industry
北京·BEIJING

内 容 简 介

本书是国家级一流本科专业（数字媒体技术）课程"游戏开发基础"的指定教材，同时被列为浙江省普通本科高校"十四五"重点立项建设教材。本书以 Unity 游戏引擎为平台，详细阐述了 12 款经典游戏的实现方法，这些游戏包括《打地鼠》、《记忆翻牌》、《拼图》、《推箱子》、《俄罗斯方块》、《华容道》、《连连看》、《三消》、《扫雷》、《贪吃蛇》、《五子棋》和《跳棋》。通过本书的学习和实践，读者能够全面理解和掌握游戏开发的核心算法，同时深刻领悟数据结构等基础知识在游戏开发中的实际应用，培养全面的工程思维。通过学习本书内容，读者不仅能够熟练掌握游戏引擎的运用，更能够灵活运用数据结构、数学思路等理论知识进行游戏创作。

本书既适合作为各院校相关专业的教材，又适合作为对游戏开发感兴趣的读者的参考书。

未经许可，不得以任何方式复制或抄袭本书之部分或全部内容。
版权所有，侵权必究。

图书在版编目（CIP）数据

游戏开发基础：基于 Unity 游戏引擎 / 张帆主编. -- 北京：电子工业出版社，2024.8
ISBN 978-7-121-47734-8

Ⅰ. ①游… Ⅱ. ①张… Ⅲ. ①游戏程序－程序设计 Ⅳ. ①TP317.6

中国国家版本馆 CIP 数据核字(2024)第 079929 号

责任编辑：戴晨辰
印   刷：三河市华成印务有限公司
装   订：三河市华成印务有限公司
出版发行：电子工业出版社
          北京市海淀区万寿路 173 信箱   邮编：100036
开   本：787×1092 1/16   印张：21.25 字数：572 千字
版   次：2024 年 8 月第 1 版
印   次：2025 年 3 月第 2 次印刷
定   价：69.90 元

凡所购买电子工业出版社图书有缺损问题，请向购买书店调换。若书店售缺，请与本社发行部联系，联系及邮购电话：(010) 88254888，88258888。
质量投诉请发邮件至 zlts@phei.com.cn，盗版侵权举报请发邮件至 dbqq@phei.com.cn。
本书咨询联系方式：dcc@phei.com.cn。

作者简介

张 帆

浙江传媒学院数字媒体技术专业主任，数字媒体与网络工程系副主任（主持工作），教师。澳门科技大学人文艺术学院博士。

Unity 价值专家，Unity 全球认证讲师。

中国大学生计算机设计大赛省级赛和国赛评委，全国高等院校计算机基础教育研究会文科专委会委员，浙江省电子竞技协会副秘书长，中国文化管理协会电子竞技工作委员会委员，全国高等院校计算机基础教育研究会网络科技与智能媒体设计专业委员会委员。

主讲课程"游戏开发基础"获浙江省本科院校"互联网+教学"示范课堂；配套大规模在线开放课程"一刻钟学会：游戏开发基础"（"中国大学 MOOC"平台）获浙江省省级线上一流课程，并被 Unity 官方中文课堂收录。《游戏开发基础——基于 Unity 游戏引擎》被列为浙江省普通本科高校"十四五"重点立项建设教材。

曾获第三届全国高校数字创意教学技能大赛一等奖，浙江传媒学院"教学十佳"，"课程思政"微课比赛一等奖，以及智慧精彩一堂课一等奖等。

主要研究方向：数字娱乐技术、人机交互技术、技术美术和人工智能等。

【序論】

前言

2018年8月21日—22日，全国宣传思想工作会议在北京召开，习近平指出："要推动文化产业高质量发展，健全现代文化产业体系和市场体系，推动各类文化市场主体发展壮大，培育新型文化业态和文化消费模式，以高质量文化供给增强人们的文化获得感、幸福感。"党的二十大报告指出："实施国家文化数字化战略，健全现代公共文化服务体系，创新实施文化惠民工程。""增强中华文明传播力影响力。坚守中华文化立场，提炼展示中华文明的精神标识和文化精髓，加快构建中国话语和中国叙事体系，讲好中国故事、传播好中国声音，展现可信、可爱、可敬的中国形象。加强国际传播能力建设，全面提升国际传播效能，形成同我国综合国力和国际地位相匹配的国际话语权。深化文明交流互鉴，推动中华文化更好走向世界。"

在这样的背景下，数字游戏产业作为数字文化经济的重要组成部分，也作为一种新的文化传播媒介，迎来了更加广阔的发展前景。随着数字游戏产业的蓬勃发展，对该领域的人才需求也提出了更高的要求。数字游戏产业不仅需要具备专业技能的设计师和开发者，还需要他们传播积极向上的社会价值观，展现精彩纷呈的中国故事。这既是一项挑战，又是一项崭新的课题。我们应该充分认识到数字游戏产业在文化发展中的重要作用，积极推动其朝着更加健康、积极的方向发展。同时，也需要加强人才培养和引进，为数字游戏产业的持续繁荣和发展提供有力支撑。

在充满神秘气息、深邃而广袤无垠的游戏世界中，每个人仿佛都是掌握命运的创造者。我们可以在这个奇幻的世界中随心所欲地探索、发现和创新，从轻松愉快的《打地鼠》到充满中国文化的《华容道》，到引人入胜的《贪吃蛇》，再到震撼人心的《三消》等，每款游戏的诞生都源自设计师无尽的智慧和辛勤的付出。这些游戏不仅为我们提供了娱乐和放松的空间，更成为传播文化、传递价值观的重要载体。

在这些游戏的背后，隐藏着无数令人着迷的细节和精心设计的谜题，那些独特的情节设定、扣人心弦的关卡挑战和别具一格的游戏设定，都为我们带来了一次又一次的惊喜。从精美的图像设计到引人入胜的音效，每个元素都经过了精心的雕琢，将我们带入一个充满魔力的游戏世界。

你是否曾对这些游戏背后的故事感到好奇，想要一探究竟呢？那些别出心裁的设计理念和奇妙的算法逻辑为游戏的成功奠定了坚实的基础。正是这些不为人知的细节和谜题，吸引着我们去探索游戏世界的每个角落，去寻找那些隐藏在游戏世界中的宝藏和秘密。在充满无限可能的游戏世界中，每个人都可以成为真正的创造者。让我们一起揭开这些游戏的神秘面纱，去领略隐藏在背后的美丽与奇迹吧！

本书将带领读者掌握游戏开发的全过程，即从一纸空白到最终实现充满活力的游戏世界的每个步骤。通过学习这门课程，读者可以了解游戏开发的核心技术和基本流程。

本书以 12 款经典游戏的算法为基础，使用 Unity 游戏引擎，引导读者实现多种类型的游戏。无论是轻松休闲的《打地鼠》和《俄罗斯方块》，还是富有策略性的《推箱子》和《贪吃蛇》，抑或是富有挑战性的《华容道》、《五子棋》和《跳棋》，读者都可以运用编程语言和游戏引擎工具实现创意，将想象力和创新转化为屏幕上活灵活现的游戏。

读者可以在实践中掌握各种基本的数据结构和算法在游戏开发中的应用，深入理解游戏运行的机制和原理。同时，读者可以意识到游戏开发并不是一个孤立的领域，而是需要融合多种学科知识，如计算机科学、数学和物理等。这也是课程设置中强调的跨学科学习的重要性，只有掌握多方面的知识，才能在这个领域中走得更远。

本书采用 Unity 游戏引擎的初衷在于它能够让游戏开发者更加关注游戏逻辑，而不是底层的技术细节，从而能够更加聚焦算法本身。

通过学习本书内容，读者不仅可以提升编程技能，还可以拓宽视野，激发创新思维，养成整体工程观念。通过不断地学习和实践，读者能够熟练使用游戏引擎，自主创作自己的游戏世界。无论是在游戏中展示独特创意，还是在设计中体现技术实力，都将是游戏开发者开发游戏的重要里程碑。

在这个旅程中，让我们一起打开游戏开发的大门，迈向创意无限的未来！无论你是初入游戏开发领域的新手，还是对游戏设计充满热情的玩家，抑或是想要拓展技能的专业人士，本书都将为你提供深入而富有挑战性的学习体验。

本书包含配套教学资源，读者可登录华信教育资源网（www.hxedu.com.cn）免费下载。读者还可借助"中国大学 MOOC"平台，搜索并参加"一刻钟学会：游戏开发基础"的在线学习课程。该线上内容不仅与本书内容相互参照，还融入了更详尽的内容，可以为读者提供更加全面的知识体系。另外，读者还可以在线提出疑问，本书的编写团队将会对每个问题予以详细解答，确保学习过程顺畅与准确。

衷心感谢浙江传媒学院媒体工程学院的同仁，尤其是谢昊老师和吉娜烨老师给予的大力协助。另外，特别感谢褚康、邓乔印、李婷、沈皓、邵嘉荣、汤玥、王杰、王子权和薛润婧 9 位优秀学生，他们为本书提供了非常重要的帮助，协助整理了大量资料，并承担了精细的校稿工作和关键的工程验证任务。感谢浙江传媒学院与网易公司共建的"网易产业学院"及与杭州电魂网络科技有限公司合办的"数字娱乐燚工作室"的鼎力支持。

由于作者水平有限，书中难免存在不足之处，欢迎广大读者批评指正。

张 帆
2024 年 8 月

目录

第1章 游戏引擎介绍 ... 1
1.1 多平台发布的 Unity ... 1
1.2 Unity 的下载和安装 ... 1
1.2.1 下载 Unity ... 2
1.2.2 安装 Unity ... 2
1.2.3 注册 Unity ... 4
1.2.4 启动 Unity ... 4
1.3 Unity 编辑器的布局 ... 5
1.3.1 标题栏 ... 5
1.3.2 主菜单栏 ... 6
1.3.3 Project（项目资源）窗口 ... 13
1.3.4 Hierarchy（层级）窗口 ... 19
1.3.5 Scene（场景）窗口 ... 21
1.3.6 Inspector（组件）窗口 ... 30
1.3.7 Game（游戏预览）窗口 ... 35
1.3.8 Console（控制台）窗口 ... 36
1.4 自定义窗口布局 ... 37
1.4.1 使用 Unity 内置的窗口布局功能 ... 37
1.4.2 自由设定窗口布局 ... 39
1.5 本章小结 ... 40
1.6 练习题 ... 40

第2章 打地鼠 ... 41
2.1 游戏简介 ... 41
2.2 游戏规则 ... 41
2.3 游戏开发核心思路 ... 41
2.3.1 洞口位置的存储结构和计算 ... 41
2.3.2 地鼠出现频率的控制 ... 42
2.3.3 地鼠的随机生成和销毁 ... 42
2.3.4 游戏时长和分数 ... 42
2.3.5 游戏流程图 ... 43
2.4 游戏实现 ... 43
2.4.1 资源的导入与 Sorting Layer ... 43
2.4.2 生成洞口 ... 45
2.4.3 地鼠的生成 ... 46
2.4.4 打击地鼠 ... 49
2.4.5 计时功能 ... 50
2.4.6 计分功能 ... 52
2.4.7 游戏结束 ... 53
2.4.8 修改地鼠出现频率 ... 53
2.4.9 修改图标 ... 54
2.4.10 添加音效 ... 55
2.4.11 重新开始游戏 ... 56
2.4.12 导出游戏 ... 56
2.5 本章小结 ... 57
2.6 练习题 ... 57

第3章 记忆翻牌 ... 58
3.1 游戏简介 ... 58
3.2 游戏规则 ... 58
3.3 游戏开发核心思路 ... 58
3.3.1 生成卡牌池 ... 58
3.3.2 洗牌 ... 59
3.3.3 卡牌状态 ... 60
3.3.4 游戏计分 ... 60
3.3.5 游戏流程图 ... 61
3.4 游戏实现 ... 61

3.4.1 游戏资源的导入与场景搭建 61
3.4.2 卡牌的位置排列 62
3.4.3 洗牌功能的实现 63
3.4.4 显示卡牌的背面与卡牌的图案 64
3.4.5 设置卡牌状态 66
3.4.6 卡牌图案的配对 67
3.4.7 计算分数与步数 69
3.5 本章小结 72
3.6 练习题 72

第4章 拼图 73

4.1 游戏简介 73
4.2 游戏规则 73
4.3 游戏开发核心思路 73
    4.3.1 碎片对象的生成与位置初始化 73
    4.3.2 原图与碎片位置的对应关系 73
    4.3.3 拖动碎片 74
    4.3.4 碎片放置位置正确性的判断 75
    4.3.5 游戏流程图 76
4.4 游戏实现 76
    4.4.1 新建工程与导入资源 76
    4.4.2 批量读取碎片 78
    4.4.3 生成碎片对象 79
    4.4.4 初始化碎片位置 79
    4.4.5 计算每个碎片的目标位置 80
    4.4.6 拖动鼠标 81
    4.4.7 判断游戏是否胜利 86
4.5 本章小结 87
4.6 练习题 87

第5章 推箱子 88

5.1 游戏简介 88
5.2 游戏规则 88
5.3 游戏开发核心思路 89
    5.3.1 地图的生成 89
    5.3.2 角色位置及移动方向与数组的对应关系 89
    5.3.3 分析角色可移动和不可移动情况 90
    5.3.4 角色移动的代码逻辑 90
    5.3.5 游戏胜利的判定 92
    5.3.6 游戏流程图 92
5.4 游戏实现 92
    5.4.1 资源的导入与切割 92
    5.4.2 角色动画的制作 94
    5.4.3 地图生成的程序实现与代码重构 95
    5.4.4 角色移动方向位置上的Tile类型检测与实现 97
    5.4.5 角色可移动情况的代码框架实现 99
    5.4.6 角色在场景中的移动功能 101
    5.4.7 打印地图快照信息 102
    5.4.8 角色移动与地图快照信息的实时更新 104
    5.4.9 推动箱子的功能实现 104
    5.4.10 移动代码的重构优化 109
    5.4.11 游戏胜利条件判断 112
5.5 本章小结 114
5.6 练习题 114

第6章 俄罗斯方块 115

6.1 游戏简介 115
6.2 游戏规则 115
6.3 游戏开发核心思路 116
    6.3.1 初始化地图 116
    6.3.2 方块类（Block）的编写 117
    6.3.3 初始化方块——方块在地图中的随机生成 117
    6.3.4 移动和旋转方块——修改位置坐标 117

6.3.5 移动和旋转方块——边界
判断 117
6.3.6 移动方块——向下
移动 118
6.3.7 消除满行方块 118
6.3.8 游戏结束判断 118
6.3.9 附加功能——提示下
一个方块 118
6.3.10 游戏流程图 118
6.4 游戏实现 119
6.4.1 资源的导入 119
6.4.2 地图的初始化与显示 121
6.4.3 地图快照信息的显示 122
6.4.4 方块的初始化 124
6.4.5 方块的随机生成 128
6.4.6 方块的移动和旋转 130
6.4.7 方块的移动、旋转与
地图的刷新 134
6.4.8 边界的判断 138
6.4.9 满行的消除 141
6.4.10 游戏结束的判断 142
6.4.11 附加功能——控制方块
自动下落的速度 144
6.4.12 附加功能——显示下
一个方块的类型 145
6.4.13 附加功能——优化重复
代码 147
6.5 本章小结 149
6.6 练习题 149

第7章 华容道 150
7.1 游戏简介 150
7.2 游戏规则 150
7.3 游戏开发核心思路 151
7.3.1 棋子类（Chess）的
编写 151
7.3.2 棋盘盘面布局的生成 151
7.3.3 移动棋子的实现 151
7.3.4 游戏胜利的判断 156
7.3.5 游戏流程图 156

7.4 游戏实现 157
7.4.1 资源的导入 157
7.4.2 棋子类的编写 158
7.4.3 棋盘布局的生成 159
7.4.4 移动棋子——选择
棋子 163
7.4.5 移动棋子——移动
1×1 棋子 169
7.4.6 移动棋子——移动
1×2 棋子 173
7.4.7 移动棋子——移动
2×1 棋子 177
7.4.8 移动棋子——移动
2×2 棋子 181
7.4.9 游戏胜利的判断 184
7.5 本章小结 188
7.6 练习题 188

第8章 连连看 189
8.1 游戏简介 189
8.2 游戏规则 189
8.3 游戏开发核心思路 190
8.3.1 编写 Tile 类 190
8.3.2 初始化地图 190
8.3.3 消除操作 191
8.3.4 绘制连接线 193
8.4 游戏实现 194
8.4.1 导入资源 194
8.4.2 编写 Tile 类 196
8.4.3 初始化地图 197
8.4.4 选择 Tile 对象 202
8.4.5 连接判断——直连、
二连、三连 205
8.4.6 连接线的绘制 209
8.5 本章小结 214
8.6 练习题 214

第9章 三消 215
9.1 游戏简介 215
9.2 游戏规则 215

9.3 游戏开发核心思路 216
    9.3.1 编写 Gemstone 类 216
    9.3.2 初始化游戏场景和地图 216
    9.3.3 消除检测 217
    9.3.4 消除宝石 218
    9.3.5 重新洗牌 219
    9.3.6 游戏流程图 219
9.4 游戏实现 220
    9.4.1 导入资源与设置场景 220
    9.4.2 编写 Gemstone 类 221
    9.4.3 初始化地图数据结构和场景 223
    9.4.4 选择宝石 225
    9.4.5 消除宝石 228
    9.4.6 重新洗牌操作 233
9.5 本章小结 235
9.6 练习题 235

第 10 章 扫雷 236

10.1 游戏简介 236
10.2 游戏规则 236
    10.2.1 《扫雷》游戏的布局 ... 236
    10.2.2 《扫雷》游戏的基本操作 237
10.3 游戏开发核心思路 237
    10.3.1 初始化游戏地图 237
    10.3.2 编写 Tile 类 237
    10.3.3 随机生成地雷 238
    10.3.4 编写辅助函数，显示地图信息 238
    10.3.5 Tile 的交互 238
    10.3.6 方格的单击（左键、遍历翻开、右键） 239
    10.3.7 游戏结束判断 239
    10.3.8 UI 控制 239
    10.3.9 游戏流程图 240
10.4 游戏实现 240
    10.4.1 资源导入与场景设置 ... 240
    10.4.2 初始化场景 241
    10.4.3 左、右键操作 248
    10.4.4 游戏结束判断 252
    10.4.5 UI 控制 256
    10.4.6 重新开始一局游戏 269
10.5 本章小结 270
10.6 练习题 270

第 11 章 贪吃蛇 271

11.1 游戏简介 271
11.2 游戏规则 271
11.3 游戏开发核心思路 271
    11.3.1 地图的生成 271
    11.3.2 食物的出现 272
    11.3.3 蛇的数据结构 272
    11.3.4 贪吃蛇的移动算法 272
    11.3.5 判断蛇头是否撞到了自身 272
    11.3.6 边界的判断 273
    11.3.7 游戏流程图 273
11.4 游戏实现 273
    11.4.1 资源导入和场景设置 ... 273
    11.4.2 编写 Node 类 275
    11.4.3 初始化场景 276
    11.4.4 贪吃蛇的移动 280
    11.4.5 结束判定 285
11.5 本章小结 286
11.6 练习题 286

第 12 章 五子棋 287

12.1 游戏简介 287
12.2 游戏规则 287
    12.2.1 《五子棋》游戏的棋盘和棋子 287
    12.2.2 《五子棋》游戏的基本规则 288
    12.2.3 落子顺序 288
12.3 游戏开发核心思路 288
    12.3.1 绘制棋盘 288
    12.3.2 绘制棋子 289
    12.3.3 落子 289

12.3.4 获胜规则 289
12.3.5 判断黑方禁手功能 290
12.3.6 游戏流程图 290
12.4 游戏实现 291
12.4.1 前期准备 291
12.4.2 初始化棋盘地图 292
12.4.3 编写落子框架 294
12.4.4 落子 296
12.4.5 连子判断和胜负判断 ... 297
12.5 本章小结 302
12.6 练习题 302

第 13 章 跳棋 303
13.1 游戏简介 303
13.2 游戏规则 303
13.3 游戏开发核心思路 304
13.3.1 棋盘排列 304
13.3.2 生成棋子 305
13.3.3 棋子在 Unity 中笛卡儿坐标系下的位置映射 ... 305
13.3.4 计算可移动位置 306
13.3.5 回合限制 307
13.3.6 游戏胜负判断 307
13.3.7 游戏流程图 307
13.4 游戏实现 308
13.4.1 前期准备 308
13.4.2 创建棋格 310
13.4.3 创建棋子 312
13.4.4 选择棋子 314
13.4.5 判断邻近可走棋格 316
13.4.6 单步移动棋子 319
13.4.7 跳子 320
13.4.8 回合限制 322
13.4.9 胜负判断 324
13.5 本章小结 327
13.6 练习题 327

第 1 章

游戏引擎介绍

1.1 多平台发布的 Unity

Unity Technologies 公司为游戏开发者提供了一款功能非常强大的开发工具——Unity，这款工具能够帮助游戏开发者轻松地开发出各种类型的游戏及虚拟现实等互动内容。Unity 支持多平台发布，这意味着游戏开发者可以将自己的作品一键发布到多个平台上，包括 Windows、Linux、Mac、iOS、Android、Web 和 Xbox 等，从而让游戏迅速在全球范围内流行起来。这便是轻量级游戏引擎的特性，功能既强大又灵活。图 1-1 所示为 Unity 的 Logo 及使用 Unity 开发的游戏。

（a）Unity 的 Logo　　　　（b）《纪念碑谷》　　　　（c）*ShadowGun*

图 1-1　Unity 的 Logo 及使用 Unity 开发的游戏

Unity 已经从单机游戏拓展到网络游戏，这是一个充满无限可能的转变。它不再局限于 PC 端，还可以应用于移动设备、虚拟现实（Virtual Reality，VR）、增强现实（Augmented Reality，AR）和体感游戏等多个领域。它的可扩展性、易用性和性价比等备受瞩目。越来越多的游戏开发者开始选择使用 Unity 进行游戏开发，你是否也想成为其中的一员呢？

本书将以 Unity 为载体，引导读者学会如何结合数据结构等知识实现 12 款经典游戏，如《打地鼠》、《华容道》、《俄罗斯方块》和《跳棋》等。只要读者认真学习本书内容，就可以成为一名优秀的游戏开发者，创造出属于自己的游戏作品。

1.2　Unity 的下载和安装

Unity 安装程序要求使用 Unity 官方网站提供的 Unity Hub 进行安装。Unity Hub 是一款复合型工具，它的功能不仅限于安装 Unity，还涵盖了多项实用子功能，如项目安排、团队协作和版本管理等。在着手安装 Unity 之前，用户需要先在 Unity 官方网站上下载 Unity Hub 并安装。

目前，Unity 提供了 3 种不同的版本，分别为 Unity Personal、Unity Plus 和 Unity Pro。由于不同群体的需求不同，因此这些版本具有各自独特的功能和权限。一般来说，如果用户为初学者或仅用于学习和开发小型游戏，那么可以选择 Unity Personal 版本，该版本提供了基本的工具和功能，能够充分满足用户的需求。如果用户需要使用更多高级功能或用于开发大型游戏并获得一定金额的收益，那么可以选择 Unity Plus 版本或 Unity Pro 版本，这两个版本提供了更多的工具和更强大的功能，能够满足更复杂的需求。

下面以 Unity Personal 版本为例，介绍 Unity 的下载、安装、注册和启动（本书主要使用 Unity 2021 版本，不同版本的操作界面有所不同，但主要内容基本一致）。

1.2.1 下载 Unity

登录 Unity 官方网站，Unity 官方网站的主页如图 1-2 所示。

（1）在进入 Unity 官方网站后，单击右上角的"下载 Unity"按钮，进入 Unity 下载界面，如图 1-3 所示。

图 1-2　Unity 官方网站的主页　　　　图 1-3　Unity 下载界面

（2）根据计算机系统下载所需版本的 Unity，如图 1-4 所示。

图 1-4　下载所需版本的 Unity

1.2.2 安装 Unity

（1）在下载器下载完成后，双击下载器安装文件，下载并安装 Unity 各组件。在下载器安装界面启动后，进入 Unity 安装界面，单击 Next 按钮，如图 1-5 所示，进入 License Agreement

界面，勾选 I accept the terms of the License Agreement 复选框，如图 1-6 所示，单击 Next 按钮。

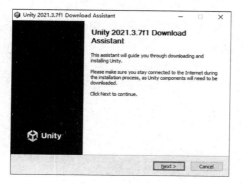

图 1-5　Unity 安装界面

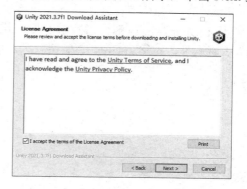

图 1-6　License Agreement 界面

（2）进入 Choose Components 界面，如图 1-7 所示。单击 Next 按钮，进入 Choose Download and Install Locations 界面，如图 1-8 所示。需要注意的是，安装路径必须是英文名称，且不要安装在带有中文名称的目录下（虽然现在 Unity 支持中文，但是还有一些不完善的地方）。如果使用的是中文 Windows 操作系统，请不要安装在桌面上，因为桌面的文件目录名为中文。

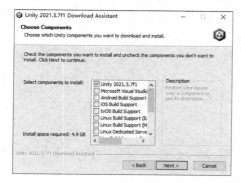

图 1-7　Choose Components 界面

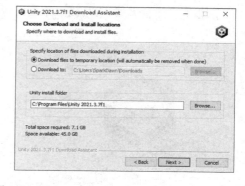

图 1-8　Choose Download and Install Locations 界面

（3）单击 Next 按钮，进入 Downloading and Installing 界面，如图 1-9 所示。安装过程比较长，请耐心等待。直到出现如图 1-10 所示的安装完成界面，单击 Finish 按钮，即可运行 Unity。

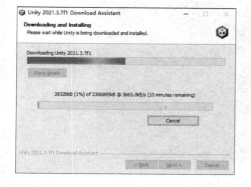

图 1-9　Downloading and Installing 界面

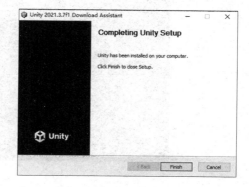

图 1-10　安装完成界面

1.2.3 注册 Unity

若已有 Unity 账号，则直接登录即可；若没有，则先单击 create one 超链接进入 Unity 官方网站注册账号，再单击 Sign In 按钮登录，如图 1-11 所示。

图 1-11　注册 Unity 账号

1.2.4 启动 Unity

可以使用以下几种方式启动 Unity：单击桌面上的 Unity 图标，如图 1-12 所示；单击左下角的"开始"按钮，在搜索框中输入 Unity，搜索 Unity，如图 1-13 所示；使用 Unity Hub，如图 1-14 所示。

图 1-12　单击桌面上的 Unity 图标

图 1-13　搜索 Unity

图 1-14　使用 Unity Hub

Unity 启动成功的界面如图 1-15 所示（为更好地展示，此处把 Unity 的 UI 设置成浅色调）。

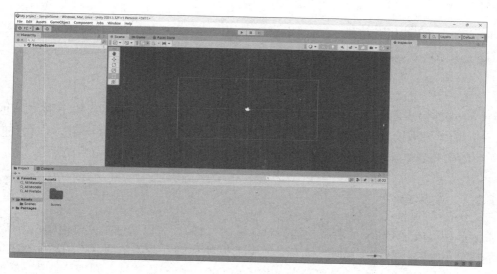

图 1-15　Unity 启动成功的界面

1.3　Unity 编辑器的布局

当第一次打开 Unity 时，显示的是默认的布局方式，如图 1-16 所示。在默认的布局方式中，显示了游戏开发中经常使用的界面窗口。当然，读者也可以根据实际需要和使用习惯重新布局。

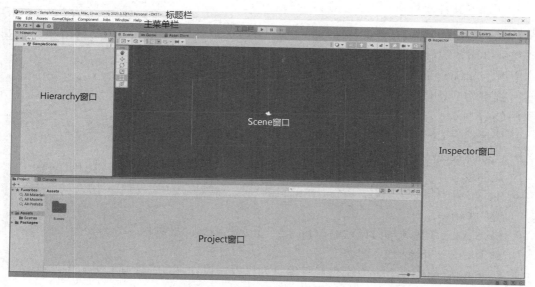

图 1-16　Unity 默认的布局方式

1.3.1　标题栏

所有的应用程序基本都有标题栏，标题栏用于显示软件和工程的基本信息，Unity 的标题栏具有同样的作用。Unity 的标题栏中显示了关于游戏工程的相关信息，如游戏场景和游戏发布平台的信息，如图 1-17 所示。

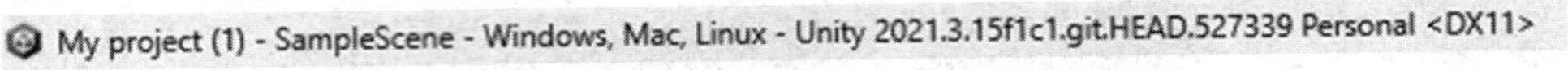

图 1-17　Unity 的标题栏

My project(1)表示工程的名称，SampleScene 表示当前打开场景的名称，Windows, Mac,Linux 表示游戏的发布平台，Unity 2021.3.15f1c1 表示软件的名称和版本。若在标题栏中添加了"*"，则表示该场景做了修改后还未保存。

1.3.2　主菜单栏

主菜单栏集成了 Unity 的所有功能菜单（见图 1-18），可以通过主菜单栏实现游戏开发。每个下拉菜单的左侧是该菜单项的名称，右侧是其快捷键。若菜单项名称的后面有省略号，则表示可以打开对应的面板；若菜单项名称的后面有一个三角符号，则表示该菜单项还有一个子菜单。若安装了其他插件，则会在主菜单栏中添加其他选项。（注意：因为 Unity 的版本更新比较快，所以各界面也许有所变动。若安装的是不同版本的 Unity，请详细查阅 Unity 官方网站。）

图 1-18　主菜单栏

1. File（文件）菜单

File 菜单用于创建、打开游戏工程和场景，以及发布游戏、关闭编辑器等，如图 1-19 所示。

图 1-19　File 菜单

- New Scene（新建场景）：创建一个新的游戏场景，快捷键为 Ctrl + N。
- Open Scene（打开场景）：打开一个已经保存的场景，快捷键为 Ctrl + O。
- Open Recent Scene（打开最近的场景）：打开最近使用的场景。
- Save（保存场景）：保存一个正在编辑的场景，快捷键为 Ctrl + S。
- Save As…（把场景另存为……）：把一个正在编辑的场景保存为另外一个场景，快捷键为 Ctrl + Shift + S。
- Save As Scene Template…：将场景保存为模板，以便可以在新建的场景中选择此模板。
- New Project…（新建工程）：创建一个新的游戏工程。
- Open Project…（打开场景）：打开一个已经存在的工程。
- Save Project（保存工程）：保存一个正在编辑的工程。
- Build Settings…（发布设置）：发布一款游戏，通过这个菜单项可以发布不同平台的游戏，快捷键为 Ctrl + Shift + B。
- Build And Run（发布并运行）：发布并运行一款游戏，快捷键为 Ctrl + B。
- Exit（退出）：退出编辑器。

2. Edit（编辑）菜单

Edit 菜单提供了回撤、复制、粘贴、运行游戏，以及设置编辑器等功能，如图 1-20 所示。

- Undo（撤销）：当误操作后，可以使用该菜单项回到上一步操作，快捷键为 Ctrl + Z。
- Redo（取消回撤）：当撤销次数过多时，可以使用该菜单项前进到上一步的撤销，快捷键为 Ctrl + Y。
- Select All（选择所有）：可以一次性选择场景中所有的对象，快捷键为 Ctrl + A。
- Deselect All（取消所有选择）：可以取消现在选择的所有对象，快捷键为 Shift + D。
- Select Children（选择子物体）：可以选择已经选择对象的子物体，快捷键为 Shift + C。
- Select Prefab Root（选择预制体）：可以选中所选择对象的预制体，快捷键为 Ctrl + Shift + R。
- Invert Selection（反选）：选择除选中的对象之外的所有物体，快捷键为 Ctrl + I。
- Cut（剪切）：选择某个对象并剪切，快捷键为 Ctrl + X。
- Copy（复制）：选择某个对象并复制，快捷键为 Ctrl + C。
- Paste（粘贴）：在剪切或复制对象后，可以把该对象粘贴到其他位置，快捷键为 Ctrl + V。
- Duplicate（复制）：复制选中的物体，快捷键为 Ctrl + D。在 Unity 中，该菜单项使用得比 Copy + Paste 更频繁。
- Rename（重命名）：对选中的物体进行重命名。
- Delete（删除）：删除某个选中的对象，快捷键为 Shift + Delete。
- Frame Selected（聚焦选择）：在选择一个物体后，使用该菜单项可以把视角移动到选中的物体上，快捷键为 F。
- Lock View to Selected（锁定视角到所选）：在选择一个物体后，使用该菜单项可以把视角移动并锁定到选中的物体上，视角会跟随所选对象的移动而移动，快捷键为 Shift +F。
- Find（查找）：可以通过在资源搜索栏中输入对象名称来查找某个对象，快捷键为 Ctrl + F。
- Play（运行）：可以运行游戏，快捷键为 Ctrl + P。
- Pause（暂停）：可以暂停正在运行的游戏，快捷键为 Ctrl + Shift + P。
- Step（逐帧运行）：可以采用一帧一帧的方式运行游戏，每选择一次，游戏运行一帧，快捷键为 Ctrl + Alt + P。
- Sign in…（登录）：登录 Unity 账号。
- Sign out（退出登录）：退出登录 Unity 账号。
- Selection（所选对象）：包括 Load Selection（载入所选）命令和 Save Selection（保存所选）命令。Load Selection 命令用于载入使用 Save Selection 命令保存的游戏对象。选择所要载入相应游戏对象的编号，便可重新选择游戏对象；Save Selection 命令用于保存当前场景中所选择的游戏对象，并赋予对应的编号，如图 1-21 所示。
- Project Settings…（工程设置）：可以根据工程的需要设置该工程的输入、音频和计时器

图 1-20　Edit 菜单

等属性，如图 1-22 所示。

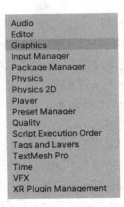

图 1-21　Save Selection　　　　图 1-22　Project Settings

- Preferences…（偏好设置）：可以设置 Unity 的外观、脚本编辑工具和 Android SDK 路径等。
- Shortcuts…（按键设置）：可以设置快捷键。
- Clear All PlayerPrefs（清除本地信息）：清除所有本地信息。
- Graphics Tier（图形层）：当前设备的图形层分类。
- Grid and Snap Settings…（网格与捕捉设置）：可以在编辑场景下对游戏对象进行移动、旋转和缩放精度的设置。

3．Assets（资源）菜单

Assets 菜单提供了对游戏资源进行管理的功能，如图 1-23 所示。

- Create（创建资源）：新建各种资源。
- Show in Explorer（打开资源所在的目录位置）：在选择某个对象后，通过操作系统的目录浏览器定位到其所在的目录中。

- Open（打开资源）：在选择某个资源后，根据资源类型以对应的方式打开。
- Delete（删除资源）：快捷键为 Delete。
- Rename（重命名资源）：重命名某个资源。
- Copy Path：复制路径，快捷键为 Alt + Ctrl + C。
- Open Scene Additive（开放场景添加剂）：将选定的场景资源的所有对象添加到当前场景中。
- View in Package Manager：在资源管理器中打开。
- Import New Asset…（导入新的资源）：通过目录浏览器导入某种需要的资源。
- Import Package（导入包）：在 Unity 中，可以通过打包的方式实现资源共享，并通过导入包来使用包资源。包资源文件的后缀是.unitypackage。

图 1-23　Assets 菜单

- Export Package…（导出包）：在编辑器中选择需要打包的资源，使用该菜单项可以把这些资源打包成一个包文件。
- Find References In Scene（在场景中找到对应的资源）：在选择某个资源后，通过该菜单项可以在游戏场景中定位到使用了该资源的对象。在使用该菜单项后，场景中没有利用该资源的对象会以黑白来显示，而使用了该资源的对象会以正常的方式显示。
- Select Dependencies（选择依赖资源）：在选择某个资源后，通过该菜单项可以显示该资源所用到的其他资源，如某个模型资源，其附属资源还包括该模型的贴图和脚本等资源。
- Refresh（刷新资源列表）：对整个资源列表进行刷新，快捷键为 Ctrl + R。
- Reimport（重新导入）：重新导入某个选中的资源。
- Reimport All（重新导入全部资源）：重新导入所有资源，如果遇到资源错乱等问题，则可以通过重新导入来尝试修复资源。
- Extract From Prefab（从 Prefab 预制体中提取）：提取模型材质。
- Update UXML Schema（UI 资源的更新）：更新 UIElements。
- Open C# Project（打开 C#工程）：打开可以编辑 C#脚本的编辑器。

4. GameObject（游戏对象）菜单

GameObject 菜单提供了创建和操作各种游戏对象的功能，如图 1-24 所示。

图 1-24　GameObject 菜单

- Create Empty（创建空游戏对象）：使用该菜单项可以创建一个只包括变换（位置、旋转和缩放）信息组件的空游戏对象，快捷键为 Ctrl + Shift + N。
- Create Empty Child（创建空的子对象）：使用该菜单项可以创建一个只包括变换（位置、旋转和缩放）信息组件的空游戏对象作为子对象，快捷键为 Alt + Shift + N。
- Create Empty Parent（创建空的父对象）：使用该菜单项可以创建一个只包括变换（位置、旋转和缩放）信息组件的空游戏对象作为父对象，快捷键为 Ctrl + Shift + G。
- 2D Object（2D 对象）：创建 2D 对象 Sprite 精灵。
- 3D Object（3D 对象）：创建 3D 对象，如立方体、球体、平面、地形和植物等，如图 1-25 所示。
- Effects（效果）：创建粒子系统或拖尾效果等。
- Light（灯光）：创建各种灯光，如点光源、平行光等，如图 1-26 所示。
- Audio（音频）：创建音频资源或音频混响区域。
- Video（视频）：创建视频资源。
- UI（用户界面）：创建 UI 上的一些元素，如文本、图片、画布、按钮和滑动条等，如图 1-27 所示。
- UI Toolkit（UI 工具）：包含用于创作 UI 的基本工具。

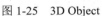

图 1-25　3D Object　　　　图 1-26　Light　　　　图 1-27　UI

- Camera（摄像机）：创建摄像机。
- Visual Scripting Scene Variables（可视化编程）：可以采用节点可视化编程方式进行逻辑设计。
- Center On Children（对齐父物体到子物体）：使父物体对齐到子物体的中心。
- Make Parent（创建父物体）：在选中多个物体后，使用该菜单项可以把选中的物体组成父子关系，其中在 Hierarchy 窗口中最上面的那个为父节点，其他为这个节点的子节点。
- Clear Parent（取消父子关系）：选择某个子物体，使用该菜单项可以取消它与父物体之间的关系。
- Set as first sibling（设置为第一个子对象）：使用该菜单项可以使选择的游戏对象在同一级中处于第一个位置，快捷键为 Ctrl + =。
- Set as last sibling（设置为最后一个子对象）：使用该菜单项可以使选择的游戏对象在同一级中处于最后一个位置，快捷键为 Ctrl + -。
- Move To View（移动到场景视图中）：在选择某个游戏对象后，使用该菜单项可以把选择的游戏对象移动到当前场景视图的中心，快捷键为 Ctrl + Alt + F。
- Align With View（对齐场景视图）：在选择某个游戏对象后，使用该菜单项可以把选择的游戏对象对齐到当前场景视图，快捷键为 Ctrl + Shift + F。
- Align View to Selected（对齐场景视图到选择的对象）：在选择某个游戏对象后，使用该菜单项可以使场景的视角对齐到选择的游戏对象上。
- Toggle Active State（切换活动状态）：使选中的游戏对象激活或失效，快捷键为 Alt + Shift + A。

5．Component（组件）菜单

使用 Component 菜单可以为游戏对象添加各种组件，如碰撞盒组件和刚体组件等。Unity 的出色之处在于以组件的软件架构来控制游戏对象，使开发游戏的流程更具有灵活性。简单来说，在 Unity 中开发游戏就是不断地为各种游戏对象添加各种组件并修改它们的属性，从而完成游戏的功能。需要注意的是，菜单会根据所添加的组件资源或插件的不同而不同，对应的菜单列表也有所变动。Component 菜单如图 1-28 所示。

- Add…（添加组件）：为选中的物体添加某个组件，快捷键为 Ctrl + Shift + A。
- Mesh（面片相关组件）：添加与面片相关的组件，如面片渲染、文字面片和面片数据。
- Effects（效果相关组件）：如粒子、拖尾效果和投影效果等。

- Physics（物理相关组件）：可以为对象添加刚体、铰链和碰撞盒等组件。
- Physics 2D（2D 物理相关组件）：可以为对象添加 2D 的刚体、铰链和碰撞盒等组件。
- Navigation（导航相关组件）：该组件模块可以用于创作寻路系统。
- Audio（音频相关组件）：可以为对象添加与音频相关的组件。
- Video（视频相关组件）：可以为对象添加与视频相关的组件。
- Rendering（渲染相关组件）：可以为对象添加与渲染相关的组件，如摄像机、天空盒等。
- Tilemap（Tile 地图编辑器相关组件）：可以为对象添加与 Tile 地图编辑器相关的组件。
- Layout（布局相关组件）：可以添加布局相关的组件，如画布、垂直布局组和水平布局组等。
- Playables（Playables 相关组件）：可以为对象添加与 Playables 相关的组件。
- AR（AR 相关组件）：可以为对象添加与 AR 相关的组件。
- Miscellaneous（杂项）：可以为对象添加动画组件、风力区域组件和网络同步组件等。
- Scripts（脚本相关组件）：可以添加 Unity 自带的或由开发者编写的脚本文件。在 Unity 中，一个脚本文件相当于一个组件，可以使用与其他组件形似的方法来控制该组件。
- UI（用户界面相关组件）：添加用户界面的相关组件，如 UI 文本、图片和按钮等。
- Event（事件相关组件）：添加与事件相关的组件，如事件系统和事件触发器等。

图 1-28　Component 菜单

6. Jobs（事务系统）菜单

Jobs 菜单提供利用 CPU 的多核和特殊的 SIMD 指令（代表单指令多数据）来尽可能有效地利用 CPU 的并行处理能力的组件。

7. Window（窗口）菜单

Window 菜单提供了与编辑器的菜单布局有关的选项，如图 1-29 所示。

- Panels（面板）：当前项目已经打开的窗口。
- Next Window（下一个窗口）：从当前的视角切换到下一个窗口。使用该菜单项，当前的视角会自动切换到下一个窗口，实现在不同的窗口视角中观察同一个物体。其快捷键为 Ctrl + Tab。
- Previous Window（前一个窗口）：将当前的操作窗口切换为编辑窗口，快捷键为 Ctrl + Shift + Tab。
- Layouts（编辑窗口布局）：可以通过它的子菜单项选择不同的窗口布局方式。
- Collaborate（Collaborate 窗口）：打开 Collaborate 窗口。

图 1-29　Window 菜单

- Plastic SCM（Plastic SCM 窗口）：打开 Plastic SCM 窗口。
- Asset Store（资源商店窗口）：打开 Unity 官方的资源商店窗口，通过该窗口可以购买所需的插件和资源。
- Package Manager（插件管理器）：打开插件管理器，可以导入插件。
- Asset Management（资源管理器）：打开资源管理器。
- TextMeshPro：打开 TextMeshPro 插件窗口。
- General（骨骼窗口）：打开骨骼窗口。
- Rendering（渲染窗口）：打开渲染窗口。
- Animation（动画编辑窗口）：打开动画编辑窗口。
- Audio（音频窗口）：打开音频窗口。
- Sequencing（序列工具窗口）：可以用于启动时间轴编辑功能。
- Analysis（分析窗口）：在 Unity 中创建应用程序时，重要的是分析和调试该应用程序，以确保其无错误，已优化并在目标设备上正确运行。
- AI（AI 窗口）：打开 AI 窗口。
- UI Toolkit（UI 工具窗口）：可以打开 UI 可视化编辑功能。

8. Help（帮助）菜单

Help 菜单提供了如当前 Unity 版本查看、许可管理和论坛地址等功能，如图 1-30 所示。

图 1-30　Help 菜单

- About Unity（关于 Unity）：选择该菜单项可以看到 Unity 当前的版本和允许发布的平台，以及创作团队等信息。
- Unity Manual（Unity 用户手册）：在选择该菜单项后，会直接连接到 Unity 官方网站的用户手册页面上。该页面主要介绍 Unity 的基本用法。
- Scripting Reference（脚本参考文档）：在选择该命令后，会直接连接到 Unity 官方网站的脚本参考文档页面上，该页面介绍了 Unity 提供的在脚本程序编写中需要用到的各种类及这些类的用法。
- Unity Learn（Unity 教程网）：可以打开 Unity 官方教程网站。
- Unity Services（Unity 服务）：在选择该菜单项后，会直接连接到 Unity 官方的服务页面上，该页面描述了 Unity 提供的用来帮助开发者制作游戏，以及吸引、留住客户并获取利润的各种服务。
- Unity Forum（Unity 论坛）：在选择该菜单项后，会直接连接到 Unity 的官方论坛上，在该论坛上可以发布各种帖子或找到一些在使用 Unity 时所遇到的问题的解决方案。
- Unity Discussions（Unity 问答论坛）：在选择该菜单项后，会直接连接到 Unity 的官方问答论坛上，在使用 Unity 时遇到的任何问题都可以通过该论坛发起提问。
- Unity Feedback（反馈页面）：在选择该菜单项后，会直接连接到 Unity 的官方反馈页面上，该页面有官方对用户的一些问题的反馈。
- Check for Updates（检查更新）：检查 Unity 是否有更新版本，如果有，就会提示用户更新。

- Download Beta…（下载测试版）：在选择该菜单项后，会直接连接到 Unity 的官方网站上，可以下载 Unity 最新的测试版。
- Manage Licenses（管理许可证）：可以更换 License 授权。
- Release Notes（发布特性一览）：在选择该菜单项后，会直接连接到 Unity 的发布特性一览页面上，该页面显示了各个版本的特性。
- Software Licenses（软件授权书）：可以打开软件的授权书信息。
- Report a Bug…（报告错误）：在使用 Unity 时，发现的引擎内在错误可以通过该窗口把错误的描述发送给 Unity 官方。
- Reset Packages to defaults（重置包到默认设置）：把 Unity 中加载的包还原到默认状态。

上面简略介绍了 Unity 的菜单功能（以上菜单可能会因为引擎版本的更新而略有不同），接下来介绍在 Unity 中使用频率较高的几个窗口。

1.3.3　Project（项目资源）窗口

Project 窗口中保存了游戏制作所需要的各种资源。常见的资源包括游戏材质、动画、字体、纹理贴图、物理材质、GUI、脚本、预制体、着色器、模型和场景文件等。可以将 Project 窗口想象成一家工厂中的原料仓库。在 Project 窗口右上角的搜索栏中可以输入要搜索资源的名称。Project 窗口如图 1-31 所示。

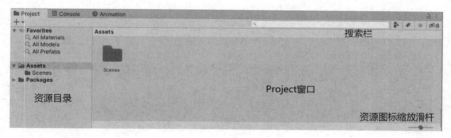

图 1-31　Project 窗口

Project 窗口的左侧是资源目录，可以通过为工程创建各种资源目录来保存不同的资源（建议为各个目录命名一个有意义的名称，这样有利于管理资源）。Project 窗口的右侧是资源目录中的具体资源，不同资源的图标是不一样的。

1. 新建资源

（1）在主菜单栏中选择 File→New Project 命令，打开 New project 窗口，如图 1-32 所示。

（2）在 Project name 文本框中输入工程名。

（3）单击 Location 文本框右侧文件夹样式的图标，打开目录浏览器，定位到要创建工程的地址，在其中新建一个目录并命名为 Chapter3-projectWindow，选择这个目录，并单击"选择文件夹"按钮，如图 1-33 所示。

（4）返回 New project 窗口，单击 Create project 按钮，一个新的工程便创建完成了（需要注意的是，最好不要把工程放在中文目录下，并且该工程的文件名不要包含中文字符）。

（5）可以选择创建 3D 工程或 2D 工程，此处选择创建 3D 工程，如图 1-34 所示。

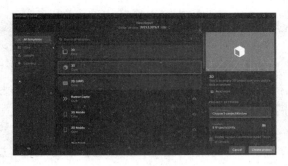
图 1-32　New project 窗口

图 1-33　单击"选择文件夹"按钮

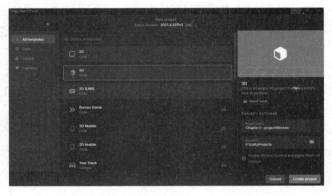
图 1-34　选择创建 3D 工程

（6）Unity 自动重启，此时编辑器是空的，如图 1-35 所示。

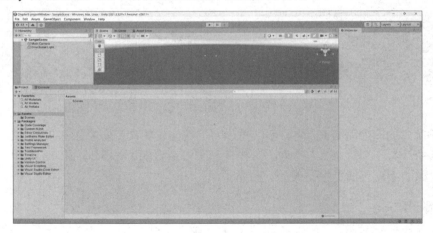
图 1-35　空的编辑器

（7）在 Project 窗口中单击鼠标右键，在弹出的快捷菜单中选择 Create→Folder 命令，此时会在 Project 窗口中生成空目录，如图 1-36 所示。

（8）在新建文件后，可以直接输入新的文件名，如果不小心单击到其他地方，就不能对文件夹名称进行修改。在名称输入错误时，可以选择该文件夹，按 F2 键就可以对它进行重命名。把新建的目录命名为_Scripts，之后这个文件夹就用来保存脚本资源，如图 1-37 所示。

（9）使用同样的方法新建文件夹_Animations（动画）、_Fonts（字体）、_Materials（材质）、_Objects（三维模型）、_Prefabs（预制体）、_Scenes（场景）、_Shaders（着色器）、_Sounds（声音）和_Textures（贴图），如图 1-38 所示。这些不同的目录将用于保存不同类型的资源。

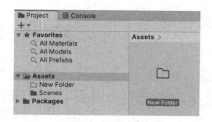

图 1-36　生成空目录　　　　　　图 1-37　新建的目录

图 1-38　最终目录结构

（10）创建子目录。先进入_Objects 目录，再进入它的子层级，使用创建目录的方法创建_Enemies（敌人模型）、_Environment（环境模型）和_Players（玩家角色模型）3 个子目录，如图 1-39 所示。当在一个目录下有子目录时，Project 窗口中的目录层级下会在对应目录的左侧出现一个三角形，该三角形表示此目录下有其他的目录，可以通过双击该目录或单击目录左侧的三角形展开该目录，如图 1-40 所示。

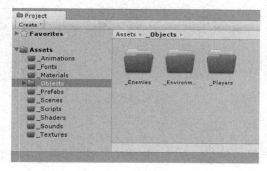

图 1-39　子目录　　　　　　　　图 1-40　展开目录

（11）先单击_Materials 目录，再单击鼠标右键，在弹出的快捷菜单中选择 Create→Material 命令，添加一个材质球，其默认名称为 New Material，如图 1-41 所示。这样便在_Materials 目录下新建了一个材质球。至于材质球应如何使用，后续会展开介绍。

图 1-41　创建材质球

通过以上步骤可以在 Project 窗口中创建新的资源。

2. 导入资源包

下面通过导入 Unity 自带的资源包来介绍如何导入已经打包的资源。

（1）在 Project 窗口中，通过单击鼠标右键打开其浮动菜单，选择 Import Unity Package→Environment 选项，此时，该资源包解压缩，并弹出一个窗口，显示的是导入的包内容，如图 1-42 所示。可以在这个窗口中选择需要的素材，或者单击 All 按钮选择全部资源，单击 None 按钮则取消所有选择，在每个资源的左侧有一个复选框，当出现符号"√"时，表示该资源被选中。若单击 Cancel 按钮，则取消该包的导入；若单击 Import 按钮，则 Unity 开始导入选中的包。需要注意的是，如果未找到该资源，那么可以前往 Unity 官方的资源商店下载。

（2）在导入 Unity 自带的资源包后，其资源都保存在 Standard Assets 目录下。可以打开 Environment 包来观察导入后的资源。如图 1-43 所示，预览导入的包的资源。

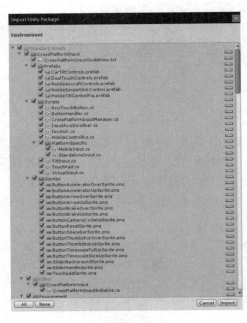

图 1-42 导入的包内容

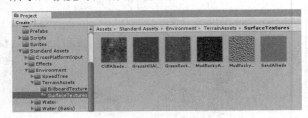

图 1-43 预览导入的包的资源

3. 导入自定义包

Unity 官方的资源商店中的很多资源都是以包的方式出售的，其文件扩展名为.unitypackage，如图 1-44 所示。当需要使用其他开发者开发的包时，可以通过导入自定义包来导入资源。

（1）在 Project 窗口中单击鼠标右键，在弹出的快捷菜单中选择 Import Package→Custom Package 命令，此时会打开文件浏览窗口，如图 1-45 所示。

图 1-44 外部包

图 1-45 文件浏览窗口

（2）展开 Resource 目录，选择其中的 iTween Visual Editor.unitypackage 文件（这个包可以用于制作补间动画的插件），单击"打开"按钮，显示 Import Unity Package 窗口，单击 Import 按钮，自定义包中的内容，如图 1-46 所示。若出现如图 1-47 所示的界面，则表示正在导入资源。

图 1-46　自定义包中的内容　　　　　　　　图 1-47　正在导入资源

（3）在导入资源后，其在 Project 窗口中所在的目录位置及其目录名称由该包来决定，如图 1-48 所示。

图 1-48　导入工程后的包资源

4．导出资源包

当需要与其他开发者共享资源时，可以将资源打包成一个资源包。下面使用 Unity 的官方例子 Tank Tutorial 工程进行介绍。

（1）在主菜单栏中选择 File→Open Project 命令，打开 Projects 界面，该界面中显示的是最近打开的工程列表，单击 Tank Tutorial 工程，如图 1-49 所示。

图 1-49　单击 Tank Tutorial 工程

（2）重新打开 Unity。假设要导出 Animators 文件夹、Materials 文件夹、Models 文件夹和 Scripts 文件夹下所有的资源，选择这 4 个文件夹（多选时可以按住 Ctrl 键并逐个选择），如图 1-50 所示。

图 1-50　选择文件夹

（3）把鼠标指针移到某个已经选中的文件夹上，单击鼠标右键，在弹出的快捷菜单中选择 Exporting package 命令，打开 Exporting package 窗口，该窗口中显示了所有需要导出的资源。可以单击 All 按钮全部选择，或者单击 None 按钮取消全部选择，或者通过直接在列表中勾选资源名称左侧的复选框来选择需要导出的素材。若勾选 Include dependencies 复选框，则表示所有被关联的资源都会被导入这个包中，一些被关联但是没有选择的资源也会被导入这个包中，如图 1-51 所示。

（4）单击 Export 按钮，在打开的窗口中可以选择需要保存该包的位置，同时输入要导出的资源包的名称，如图 1-52 所示。单击"保存"按钮，若出现如图 1-53 所示的界面，则表示已经开始导出包，并且可以查看导出的进度。

（5）如果打包完成，Unity 就会自动打开该包保存的位置，如图 1-54 所示。

图 1-51　Exporting package 窗口

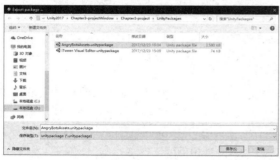

图 1-52　输入要导出的资源包的名称

图 1-53　导出包

图 1-54　打包完成

5. 使用拖动的方法导入已有的资源

Unity 允许直接在外部目录下把素材拖入 Project 窗口中，执行该操作会把素材复制到工程的 Assets 目录下的特定文件夹中。

（1）打开 Chapter3-projectWindow 工程，在 Project 窗口中选中 _Textures 目录，当前该目录下没有任何资源，如图 1-55 所示，后续需要把贴图资源拖到该目录下。

图 1-55　空的 _Textures 目录

（2）先打开操作系统的目录浏览器，再打开 Chapter3 目录下的 Textures 文件夹，外部资源素材如图 1-56 所示。

（3）直接把贴图资源拖到 Project 窗口中，这样便可以在 _Textures 目录下完成贴图资源的导入，如图 1-57 所示。

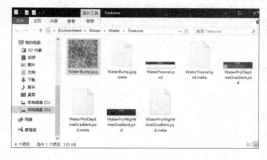

图 1-56　外部资源素材　　　　　　图 1-57　导入贴图资源

以上介绍的是 Project 窗口的基本操作，熟悉这些操作有利于提高游戏开发的工作效率。需要强调的是，要养成时刻将资源分类并整理的习惯，以便在开发过程中可以迅速找到需要的资源，尤其是当游戏非常庞大时更加重要。

1.3.4　Hierarchy（层级）窗口

Hierarchy 窗口用于保存游戏场景中存在的游戏对象，显示的内容是游戏场景中游戏对象的层次结构图。该窗口中列举的游戏对象与游戏场景中的对象是一一对应的。打开 Tank Tutorial 工程，在 Project 窗口中双击 CompleteMainScene 场景，此时，Hierarchy 窗口中就会出现对应的列表，如图 1-58 所示。其中，左侧是 Hierarchy 窗口，右侧是 Scene 窗口（该窗口将在 1.3.5 节介绍）。

在 Hierarchy 窗口中选择任意一个游戏对象，Scene 窗口中相应的游戏对象也会被选择，如选择 CompleteLevelArt 对象，Scene 窗口中相应的游戏对象也会被选上，如图 1-59 所示。

图 1-58　Hierarchy 窗口（该游戏画面可以在 Unity 官方的资源商店中搜索 Tank！Tutorial 下载）

图 1-59　通过 Hierarchy 窗口选择 Scene 窗口中的游戏对象

接下来介绍如何使用 Hierarchy 窗口创建一个游戏对象。

在 Hierarchy 窗口中创建简单的游戏对象的操作步骤如下。

（1）新建一个工程并命名为 Chapter3-HierarchyWindow，此时该工程是空的。

（2）查看 Hierarchy 窗口，可以看到只有一个 Main Camera 对象，该对象是主摄像机。当选择主摄像机时，Scene 窗口的右下角会出现一个预览窗口，这个预览窗口便是主摄像机当前所看到的场景，如图 1-60 所示。

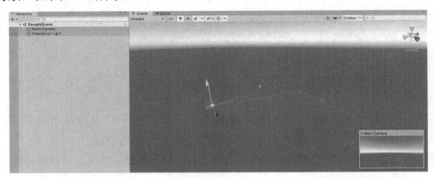

图 1-60　预览窗口

（3）在 Hierarchy 窗口中，单击左上角的图标 ➕，弹出的浮动菜单栏如图 1-61 所示，该浮动菜单栏与主菜单栏中的 GameObject 菜单上的部分菜单项大体上是一致的。

（4）选择 3D Object→Cube 命令，在 Scene 窗口中创建一个立方体，如图 1-62 所示。

 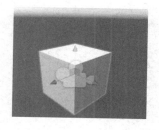

图 1-61　对象创建菜单栏　　　　　图 1-62　创建一个立方体

在 Hierarchy 窗口中创建的 Cube 对象应该是资源的一种，为什么在 Project 窗口中没有显示呢？在 Unity 中，有一些简单的对象属于内置的资源，可以直接通过 Hierarchy 窗口或 GameObject 菜单来创建（如摄像机、灯光和简单的几何体），无须通过 Project 窗口来创建。

1.3.5　Scene（场景）窗口

Unity 是一款所见即所得的游戏编辑软件，该软件可以通过可视化的方式对游戏场景进行编辑，从而为游戏开发者提供直观的操作方式。在 Unity 中，游戏的场景编辑都是在 Scene 窗口中完成的，在这个窗口中使用游戏对象的控制柄可以移动、旋转和缩放场景里面的游戏对象。在打开一个场景后，该场景包括的游戏对象便会显示在 Scene 窗口中，如图 1-63 所示。

图 1-63　Scene 窗口

1．场景视图控制面板

场景视图控制面板用来控制场景视图的显示方式，位于 Scene 窗口的顶端，如图 1-64 所示。

图 1-64　场景视图控制面板

下面从左到右介绍场景视图控制面板。

- Shading Mode（绘制模式）：场景视图控制面板中第一个下拉菜单的第一部分，用于设置 Scene 窗口的绘制模式，如图 1-65 所示。
  - Shaded（贴图模式）：以带有贴图的方式显示场景，如图 1-66 所示。
  - Wireframe（线框模式）：以线框的方式显示场景，如图 1-67 所示。
  - Shaded Wireframe（贴图线框模式）：以带有贴图和线框的方式显示场景，如图 1-68 所示。

图 1-65　场景视图控制面板中的第一个下拉菜单

- Miscellaneous（渲染模式）：场景视图控制面板中第一个下拉菜单的第二部分，用来设置场景视图的渲染方式。
  - Shadow Cascades（阴影级联显示模式）：显示阴影的渲染情况，如图 1-69 所示。
  - Render Paths（渲染路径模式）：采用颜色标记场景中每个对象的渲染方式，绿色代表采用"延迟灯光"渲染，黄色代表采用"前向"渲染，红色代表采用"顶点着色"渲染，如图 1-70 所示。
  - Alpha Channel（Alpha 通道模式）：采用带有 Alpha 信息的方式渲染，如图 1-71 所示。
  - Overdraw（透明轮廓模式）：采用透明轮廓和透明颜色累积的方式来表示对象被重绘的次数，如图 1-72 所示。
  - Mipmaps（Mipmaps 模式）：通过颜色标记的方式来显示理想的贴图尺寸，红色表示该贴图的尺寸大于目前需要的尺寸，蓝色表示该贴图的尺寸小于目前需要的尺寸。当然，贴图所需要的尺寸是根据游戏在运行时摄像机与物体贴图之间的距离来决定的。通过这种模式可以对贴图的尺寸进行调整，如图 1-73 所示。
  - Texture Streaming（贴图串流）：显示 Unity 加载到内存中的 Mipmap 层级。
  - Sprite Mask（精灵蒙版）：用于控制精灵的可见性。

图 1-66　Shaded

图 1-67　Wireframe

图 1-68　Shaded Wireframe

图 1-69　Shadow Cascades

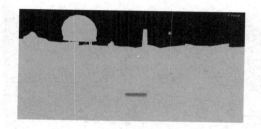

图 1-70　Render Paths

图 1-71　Alpha Channel

图 1-72　Overdraw

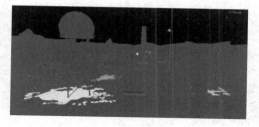

图 1-73　Mipmaps

- Deferred（设置延迟渲染）：场景视图控制面板中第一个下拉菜单的第三部分，这些模式可以用来查看渲染的每个参数的具体情况（如反光率、高光、光滑度和法线）。
- Global Illumination（全局光照）：场景视图控制面板中第一个下拉菜单的第四部分，这些模式可以用来帮助全局光照系统可视化，如 UV 图、反光率、发光、辐射、方向和烘焙等。与 Global Illumination 对应的还有 Realtime Global Illumination 和 Baked Global Illumination。Realtime Global Illumination 用于显示实时全局光照信息，Baked Global Illumination 用于显示烘焙全局光照信息。
- Material Validation（材质验证）：可以验证材质的相关信息。

绘制和渲染模式按钮后面有 6 个按钮，用于设置场景，如图 1-74 所示。

图 1-74　用于设置场景的 6 个按钮

第一个按钮用于切换场景的 2D 视图和 3D 视图；第二个按钮用于控制采用默认的灯光照明，还是采用场景中已有的灯光照明；第三个按钮用于控制是否在 Scene 窗口中播放音频；第四个按钮用于控制是否显示天空盒、雾效等；第五个按钮用于控制是否显示某类游戏对象；第六个按钮用于控制如何显示辅助网格。

在场景中，灯光、摄像机、碰撞盒和音源等都会以辅助图标标记出来，以便对这些对象进行控制，但在最终的游戏画面中这些图标是不显示的。可以通过辅助图标设置（Gizmos）面板来控制是否显示这些图标，以及修改图标的尺寸，如图 1-75 所示。

3D Icons 用于控制是否显示辅助图标，右侧的滑动杆用于控制所有辅助图标的尺寸，如图 1-76 所示。

Show Grid：用于控制是否显示网格，图 1-77 所示为取消网格显示的效果。

图 1-75　辅助图标设置面板

图 1-76 设置辅助图标

图 1-77 取消网格显示的效果

其他辅助图标可以用于控制是否显示某种特定类型的辅助图标。

2．视图变换控制

图 1-78 视图控制手柄

在 Scene 窗口中，右上角有一个视图变换控制图标，该图标为视图控制手柄，用于切换场景的视图角度，如自顶向下、自左向右、透视模式、正交模式等，如图 1-78 所示。

视图变换控制图标有 6 个坐标手柄，以及位于中心的透视控制手柄。单击 6 个坐标手柄中的一个，可以把场景切换到对应的视图中，而单击中心的立方体或下方的文字标记可以切换为正交模式与透视模式。图 1-79～图 1-83 所示分别为右视图、前视图、顶视图、投影模式（近大远小）和正交模式（无近大远小效果）。

图 1-79 右视图

图 1-80 前视图

图 1-81 顶视图

图 1-82 投影模式（近大远小）

图 1-83 正交模式（无近大远小效果）

在某种视图模式下，当视图变换控制图标的下方已经标注了该视图的名称时，视图名称的左侧会有一个表示是否正交或透视显示的小图标，如图1-84和图1-85所示。可以通过单击视图名称切换透视模式与正交模式。

3．场景视图导航

使用场景视图导航可以让场景搭建变得更加便捷和高效。场景视图导航主要采用快捷键的方式来控制，并且Unity编辑器主功能面板上的图标会显示出当前的操作方式。场景视图中显示的当前导航图标如图1-86所示。

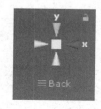 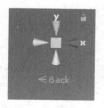 

图1-84　正交显示　　　　图1-85　透视显示　　　　图1-86　场景视图中显示的当前导航图标

- Arrow Movement：采用键盘上的方向键来实现场景漫游。单击Scene窗口就可以将其激活，使用方向键↑和方向键↓可以控制摄像机向前和向后移动，使用方向键←和方向键→可以控制摄像机向左和向右移动。配合使用Shift键可以让移动加快。
- Focus：聚焦定位。在Scene窗口或Hierarchy窗口中选择某个物体，按F键，可以使视图聚焦到该物体上。
- 移动视图：使用Alt键+鼠标滚轮或直接使用鼠标左键，可以对场景视图中的摄像机进行平移。如果处于场景编辑状态下，那么可以使用Q键来切换到场景导航操作。其图标为一个手形形状。
- 缩放视图：使用Alt键+鼠标右键或直接使用鼠标滚轮，可以对场景视图中的摄像机执行推拉操作。其图标为一个放大镜🔍。
- 旋转视图：使用Alt键+鼠标左键或直接使用鼠标右键，可以对场景视图中的摄像机进行旋转。其图标是一个眼睛👁。
- 飞行穿越模式：使用键盘上的W键、A键、S键和D键+鼠标右键，可以对场景视图中的摄像机进行移动和旋转，配合鼠标滚轮可以控制摄像机移动的速度。

4．场景对象的编辑

健全的游戏引擎编辑器一般都是通过"所见即所得"的方式来编辑场景的。场景的编辑可以通过移动、旋转和缩放物体来操作。在编辑器的左上角有一排按钮，这排按钮用于对游戏对象进行移动、旋转和缩放，如图1-87所示。

图1-87　场景对象控制按钮

第一个按钮是场景视图操作，前面已经讲解过，此处不再赘述；第二个按钮为对象移动按钮，可以对场景中的对象进行平移，快捷键是W键；第三个按钮为旋转按钮，可以对对象进行旋转，快捷键是E键；第四个按钮为缩放按钮，可以对对象进行缩放操作，快捷键是R键；第五个按钮为矩形变换按钮，可以对对象进行缩放、旋转操作，多用于UI元素，快捷键是T键；第六个按钮为对象操作的控制设置按钮。每个被操作的对象上都有对应的操作柄，每个操作柄都有相应的轴向控制柄，以便在视图中对它进行操作。图1-88～图1-91所示分别为移动图标、旋转图标、缩放图标和矩形变换图标。

  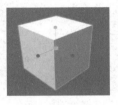 

图 1-88　移动图标　　　图 1-89　旋转图标　　　图 1-90　缩放图标　　　图 1-91　矩形变换图标

接下来介绍如何在 Unity 中对场景进行编辑。

1）场景编辑

（1）打开 Unity，新建一个工程并命名为 Chapter3-SceneEdit。

（2）在 Hierarchy 窗口中单击图标➕，选择 3D Object→Plane 命令，创建一个平面，如图 1-92 所示。

（3）在 Scene 窗口中选中创建的平面，按 F 键，使该平面位于视图的中心，同时使视口中心对准平面，如图 1-93 所示。

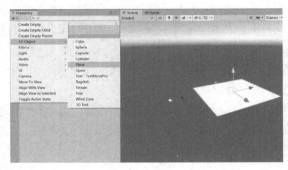 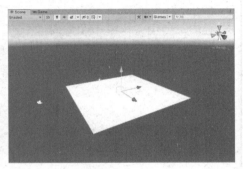

图 1-92　创建一个平面　　　　　　　　　　图 1-93　使视口中心对准平面

（4）在 Hierarchy 窗口中单击图标➕，选择 3D Object→Cube 命令，创建一个立方体，如图 1-94 所示。

（5）在 Scene 窗口选中创建的立方体，如果比较难选中，则先通过 Hierarchy 窗口选择 Cube 命令，接着按 F 键，使 Scene 窗口中的摄像机聚焦到立方体上（如果使用 Hierarchy 窗口来选择对象，则在按 F 键之前，先单击 Scene 窗口，以激活该窗口），如图 1-95 所示。

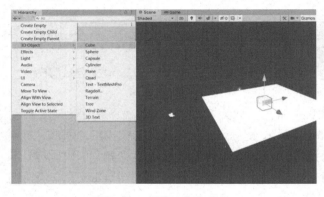 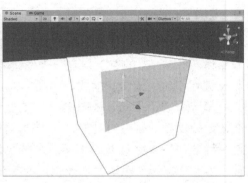

图 1-94　创建一个立方体　　　　　　　　　　图 1-95　使视口中心对准立方体

(6)按 W 键,切换到对象移动操作上,选择 Y 轴方向的操作柄,按住鼠标左键,将立方体向上拖动,使立方体在平面上方。移动操作柄共有 3 个：X 轴向,相对于对象的左右方向,用红色来表示；Y 轴向,相对于对象的上下方向,用绿色来表示；Z 轴向,相对于对象的前后方向,用蓝色来表示。当激活某个操作柄时,该操作柄会变成黄色。在移动操作柄时,如果想在由两个轴向定义的平面内移动,则可以选择该操作柄中心附近的操作平面（在平移操作模式下,按住 V 键可以捕获顶点。利用该功能可以使操作点捕获选中对象的某个点,同时移动立方体,也可以将被选中的点对齐到场景中其他对象的点上）,如图 1-96 所示。

(7)按住鼠标右键,拖动 Scene 窗口,使视口中心在立方体的左侧,如图 1-97 所示。

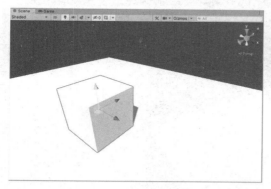 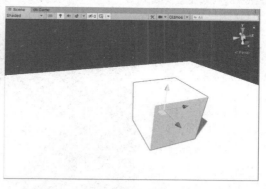

图 1-96　移动立方体　　　　　　　　　图 1-97　改变视口位置

(8)在 Hierarchy 窗口中单击图标,选择 3D Object→Cylinder 命令,创建一个圆柱体,如图 1-98 所示。此时会发现,在创建对象时,对象的位置是根据视口中心来确定的。

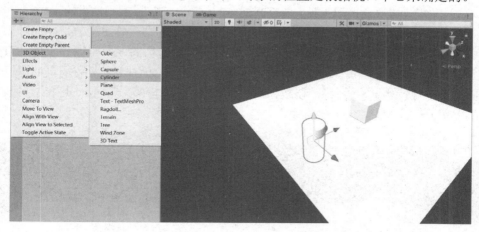

图 1-98　创建一个圆柱体

(9)使用移动工具调整圆柱体的位置,如图 1-99 所示。

(10)按 E 键,把对象操作工具切换为旋转操作,如图 1-100 所示。与移动工具相似,绕 X 轴旋转的操作环为红色,绕 Y 轴旋转的操作环为绿色,绕 Z 轴旋转的操作环为蓝色,绕视图视线方向旋转的操作柄的外圈为灰色。

(11)选择绕 Z 轴旋转的操作环,被激活的操作环会以黄色高亮显示,按住鼠标左键拖动,将圆柱体旋转 90°左右,如图 1-101 所示。

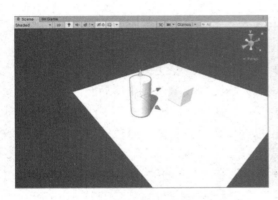
图 1-99　调整圆柱体的位置

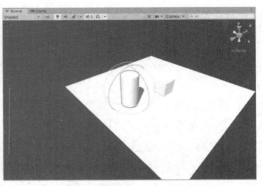
图 1-100　旋转圆柱体

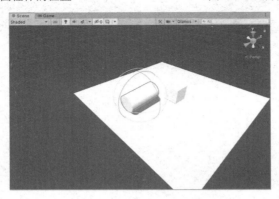
图 1-101　将圆柱体旋转 90°左右

（12）按 W 键，切换为移动工具，调整圆柱体的位置，使它接触到地下的平面，如图 1-102 和图 1-103 所示。此时会发现，操作柄的朝向发生变化。这里需要注意的是，此时操作柄的位置和朝向与该对象的局部坐标系是一致的。如果想使操作柄的朝向与世界坐标系对齐，也就是 X 轴永远对齐左右方向，Y 轴永远朝向场景上下方向，Z 轴永远对齐场景的深度方向，那么可以单击按钮 ，该按钮用于切换操作柄对齐方式，Local 表示对齐到局部坐标系，World 表示对齐到世界坐标系。

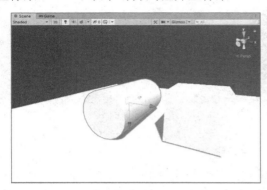
图 1-102　局部坐标系

图 1-103　世界坐标系

（13）对视口进行操作，使视口中心位于立方体的前方空白处，如图 1-104 所示。

（14）在 Hierarchy 窗口中单击图标 ✚，选择 3D Object→Sphere 命令，创建一个球体，如图 1-105 所示。

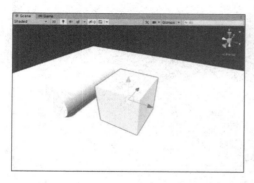

图 1-104　调整视口中心的位置

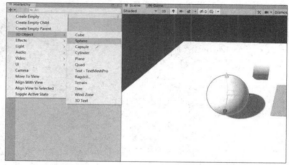

图 1-105　创建一个球体

（15）按 W 键，切换为移动工具，选中球体，调整其位置，如图 1-106 所示。

（16）选中球体，按 R 键，切换为缩放工具，如图 1-107 所示。缩放工具的轴向与移动工具的轴向相似。红色操作柄表示沿着 X 轴向缩放，绿色操作柄表示沿着 Y 轴向缩放，蓝色操作柄表示沿着 Z 轴向缩放，选择中心的操作柄可以使对象在各个轴向上等比例缩放。被选中的操作柄以黄色表示。

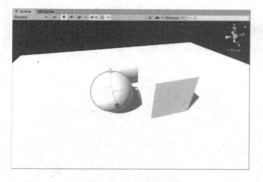

图 1-106　调整球体的位置

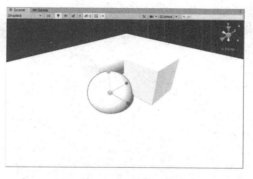

图 1-107　切换为缩放工具

（17）选择中心的操作柄，按住鼠标左键拖动，使球体的半径缩放为原来的一半，如图 1-108 所示。

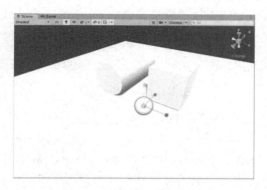

图 1-108　对球体进行缩放

（18）为场景打上灯光。在 Hierarchy 窗口中单击图标 ，选择 Light→Directional Light 命令，创建平行光（平行光在 Scene 窗口中使用图标 ※ 表示），这样整个场景便被照亮（关于灯光的用法后面会介绍，这里不展开介绍），如图 1-109 所示。

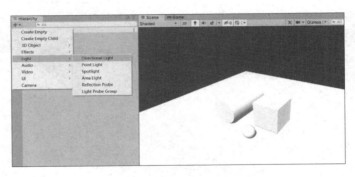

图 1-109　创建平行光

（19）选择 File→Save Scene 命令，或者直接使用快捷键 Ctrl + S 对场景进行保存。此时会弹出一个目录浏览器，在"文件名"文本框中输入 SceneEdit 作为该场景的名称，单击"保存"按钮。该步骤使场景保存在需要的目录中，此时需要注意的是，场景文件一定要放在工程目录的 Assets 目录下或该目录的子目录下，如图 1-110 所示。

（20）在场景保存完成后，Project 窗口中会出现一个关于场景的图标，如图 1-111 所示，直接单击该图标就会打开该游戏场景。

图 1-110　保存场景

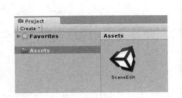

图 1-111　场景保存完成后的图标

2）控制场景编辑器窗口的显示图层

在工具栏的最右侧有一个用来控制场景编辑器窗口的显示图层按钮，如图 1-112 所示，该按钮的名称为 Layers。

图 1-112　Layers 按钮

单击 Layers 按钮会弹出一个浮动菜单，如图 1-113 所示。眼睛图标表示渲染出所有的层。如果选择 Nothing 命令，则 Scene 窗口将不显示任何内容。如果选择 Everything 命令，则 Scene 窗口将显示所有图层的内容。也可以取消选择其中的某些图层。

图 1-113　Layers 浮动菜单

1.3.6　Inspector（组件）窗口

在使用 Unity 创作游戏时，游戏的场景都是由游戏对象组成的，而游戏对象又包括模型面片、脚本和音频等组件，游戏对象的属性和行为是由添加到该游戏对象上的组件来决定的。

Unity 提供了一个添加组件和修改组件参数的窗口，即 Inspector 窗口。当选择了某个游戏对象时，在 Inspector 窗口中便会显示出已经添加到该游戏对象上的组件和这些组件的属性，如图 1-114 所示。

Inspector 窗口中所显示的游戏对象有几个固定的属性和组件，如图 1-115 所示。

图 1-114　Inspector 窗口

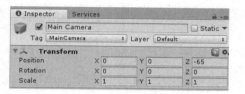

图 1-115　固定的属性和组件

1．图标设置

在 Inspector 窗口中，左上角为一个图标，用于标记不同的对象，可以根据需要对该图标进行修改。单击该图标会出现一个面板，如图 1-116 所示。使用该面板可以修改图标的形状和颜色。当单击 Other 按钮时，会出现一个贴图列表面板，如图 1-117 所示，可以通过选择自定义贴图来修改图标。

图 1-116　图标设置面板

图 1-117　贴图列表面板

2．激活复选框

复选框用于控制游戏对象在游戏场景中是否已被激活。当把"√"删除后，该游戏对象便不会在场景中显示，并且所有的组件也会失效，虽然该游戏对象仍然保留在场景中，如图 1-118 所示。

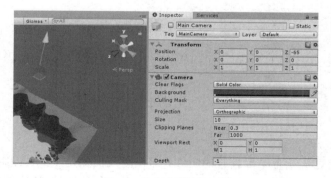

图 1-118　注销游戏对象

3．对象名称

复选框的后面是一个文本框，可以通过该文本框修改游戏对象的名称。也可以在 Hierarchy 窗口中先选择游戏对象，再按 F2 键对其名称进行修改。

4．Static 复选框

Static 复选框用于控制是否把游戏对象设置成静态的。对于场景中一些静态的游戏对象，可以勾选 Static 复选框，这样不但可以在一定程度上减少游戏渲染工作量，而且如果要对场景中的游戏对象进行光照贴图烘焙、寻路数据烘焙和 Occlusion Culling 运算，也需要将其设置成静态的。

5．Tag

Tag 用来为游戏对象加上有意义的标签名称。标签的主要作用是为游戏对象添加一个索引，这样可以为在脚本程序中使用标签寻找场景中添加了该标签的游戏对象提供便利。可以把标签想象成某类游戏对象的别名。下面以班级举例，在一个班级中有很多个学生，每个学生都有自己的姓名，当要叫某个学生打扫卫生时，可以直接喊他的名字，但如果想叫全班的学生一起过来，那么可以直接喊班级的名称，如某某班级的学生都过来打扫卫生，那么"某某班级"便是这个班级学生的标签，而每个学生的名字便是每个对象的具体名称。在 Unity 游戏场景中，可以为多个游戏对象添加一个相同的标签，在之后编写脚本程序时，可以直接寻找该标签，这样能够找到使用该标签的所有游戏对象。

6．Layer

可以先设置游戏对象的图层，再令摄像机只显示某个图层上的对象。另外，在设置图层时，可以使物理模拟引擎只对某个图层起作用。

7．Prefab

当某个游戏对象是由 Prefab 预制体生成时，便会在此处显示该操作按钮。单击 Select 按钮，在 Project 窗口中可以找到该对象所引用的 Prefab；单击 Revert All 按钮，可以对当前针对该对象所做的修改执行回撤操作，并重新引用该对象所引用的 Prefab 的原有属性；单击 Apply All 按钮，可以把对该对象的修改应用到原来的 Prefab 上，此时，所有在场景中引用了该 Prefab 的游戏对象就会同步做修改，如图 1-119 所示。

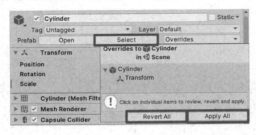

图 1-119　Prefab

8．Transform

Transform 是所有游戏对象都具有的组件，即使游戏对象是空的。该组件负责设置游戏对象在场景中的 Position（位置）、Rotation（旋转角度）和 Scale（缩放比例）。如果想精确设置某个游戏对象的变换属性，则可以直接在这个组件中修改对应的参数。当一个游戏对象没有父物体时，这些参数是相对于世界坐标系的；当一个游戏对象有父物体时，这些参数是相对于父物体的局部坐标系的。

接下来介绍通过 Transform 组件操作游戏对象的例子。

（1）打开新建的工程 Chapter3-SceneEdit，在进入游戏引擎后，通过单击打开 SceneEdit 场景文件。如果 Scene 窗口中没有物体，则可以先在 Hierarchy 窗口中任意选择一个物体，再激活 Scene 窗口，最后按 F 键，使视图定位到选择的物体上。

（2）在 Hierarchy 窗口中选择 Cube 对象，按 F2 键，将其名称修改为 Box，按 Enter 键，完成修改，如图 1-120 所示。

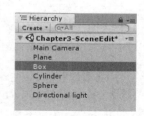

图 1-120 在 Hierarchy 窗口中修改对象名称

（3）在场景编辑器中选择圆柱体，在 Inspector 窗口中将其名称修改为 Column，按 Enter 键，完成修改，如图 1-121 所示。

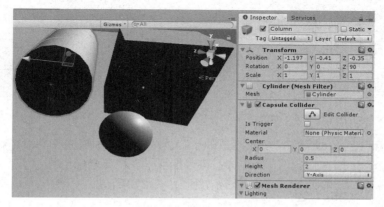

图 1-121 在 Inspector 窗口中修改对象名称

（4）继续选择圆柱体，在它的 Transform 组件中，设置 Position（位置）中的 X 为-0.17，Y 为-1，Z 为-6.5，设置 Rotation（旋转）中的 Z 为 270，如图 1-122 所示。可以看出，圆柱体的位置和旋转角度已被精确定义。

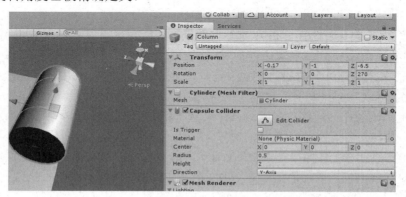

图 1-122 在 Inspector 窗口中修改圆柱体的位置和旋转角度

（5）把 Sphere 重命名为 Ball，如图 1-123 所示。由以上步骤可以看出，可以通过 Hierarchy 窗口或 Inspector 窗口来修改游戏对象的名称。

（6）选择 Ball 对象，在 Inspector 窗口中修改其位置和缩放比例，如图 1-124 所示。

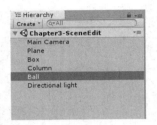

图 1-123　把 Sphere 重命名为 Ball

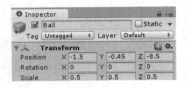

图 1-124　在 Inspector 窗口中修改 Ball 对象的位置和缩放比例

（7）把 Ball 对象作为 Column 对象的子物体。在 Hierarchy 窗口中，选择 Ball 对象并按住鼠标左键，将其拖到 Column 对象上，如图 1-125 所示。在松开鼠标左键后，Ball 对象就会成为 Column 对象的子物体，如图 1-126 所示。

图 1-125　将 Ball 对象拖到 Column 对象上　　图 1-126　Ball 对象成为 Column 对象的子物体

（8）对比 Ball 对象成为 Column 对象的子物体前的坐标和成为 Column 对象的子物体后的坐标，可以发现，现在该坐标值参考了 Column 对象的局部坐标系，如图 1-127 所示。

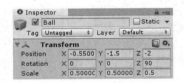

图 1-127　Ball 对象的坐标变为以 Column 对象的局部坐标系为参考

（9）选择 Column 对象，此时其子物体也会被选中，对 Column 对象进行平移、旋转和缩放操作，Ball 对象也参照父物体的变换而做相应的变换。当子物体变换时，父物体的变换并没有受到影响。此时需要注意的是，在操作父物体的变换时，子物体的变换中心点在父物体的局部坐标系的中心上。

（10）取消父子关系。选择 Ball 对象并按住鼠标左键，把该对象拖出 Column 对象，此操作取消了 Column 对象和 Ball 对象之间的父子关系。

（11）选择多个物体并同时进行变换操作。在 Scene 窗口中，按住 Ctrl 键，逐个选择 Ball 对象、Box 对象和 Column 对象，此时会发现变换操作柄将逐步移到最后选择的对象上，如图 1-128 所示，该操作柄最终决定了这几个被选择的物体的变换参考中心。当对这几个物体进行旋转时，其参考中心在最后选择的对象上；当对多个物体进行缩放时，参考中心为每个物体的局部坐标系的中心（如果是在 Hierarchy 窗口中选择多个物体，那么变换操作柄将在第一个被选择的对象上）。

（12）把多选的物体变换参考中心切换到所有被选物体的中心。在切换局部坐标系与世界坐标系按钮的左侧，有一个可以用于切换多选物体参考坐标的按钮，如图 1-129 和图 1-130 所示。Pivot 表示将单个物体的局部坐标系作为参考坐标，在单击这个按钮后会把参考中心切换到多个物体的中心上。这时再对这几个物体进行变换操作时，无论是移动、旋转还是缩放，都是以这几个物体的中心为参考进行变换的，如图 1-131 和图 1-132 所示。

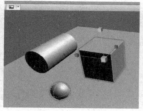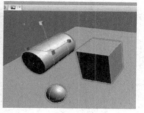

图 1-128　选择多个物体后变换操作柄的位置发生变化

图 1-129　Pivot 模式　　　　　　　　图 1-130　Center 模式

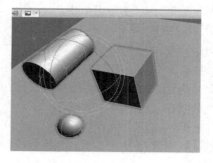

图 1-131　移动操作柄　　　　　　　　图 1-132　旋转操作柄

1.3.7　Game（游戏预览）窗口

在 Game 窗口中可以预览游戏的最终效果，如图 1-133 所示。

图 1-133　Game 窗口

Game 窗口经常搭配工具栏中的"播放"按钮和"暂停"按钮来使用，如图 1-134 所示。单击第一个按钮可以播放游戏，快捷键为 Ctrl + P；单击第二个按钮可以暂停游戏，快捷键为 Ctrl + Shift + P；单击第三个按钮可以逐帧播放游戏，快捷键为 Ctrl + Alt + P。

1）Display：选择显示器

当连接了多个显示器时，可以选择在哪个显示器上显示。

2）设置分辨率

在 Game 窗口中，左上角是分辨率设置按钮，可以根据需要设置不同的播放分辨率。单击该按钮会弹出一个浮动菜单（发布平台不同，浮动菜单会有所不同）。图 1-135 所示为 PC

平台下的分辨率设置。

图 1-134 "播放"按钮和"暂停"按钮　　　　图 1-135 PC 平台下的分辨率设置

3）Maximize on Play 按钮

在单击该按钮的情况下，单击"播放"按钮，Game 窗口会全屏显示。

4）Mute Audio 按钮

在单击该按钮的情况下，当游戏运行时不播放音频。

5）Stats 按钮

单击该按钮会出现一个与游戏运行效率有关的面板，通过这个面板可以查看目前游戏的运行效率和状态，如图 1-136 所示。

6）Gizmos 按钮

在单击该按钮的情况下，Game 窗口就会显示场景中的辅助图标，如图 1-137 所示。

图 1-136　游戏的运行效率和状态　　　　图 1-137　场景中的辅助图标

1.3.8　Console（控制台）窗口

Console 是 Unity 用于调试与观察脚本运行状态的窗口（最下方是状态栏，该栏也在 Unity 编辑器的最下方。当有信息输出时，双击状态栏的信息便可弹出 Console 窗口）。当脚本编译出现警告或错误时，可以从 Console 窗口中查看到错误的位置，以便修改。白色的文本表示普通的调试信息，黄色的文本表示警告信息，红色的文本表示错误信息。Console 窗口通常与脚本编程息息相关。Console 窗口如图 1-138 所示。

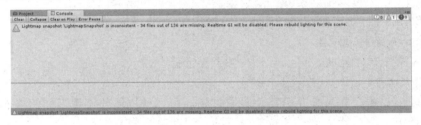

图 1-138　Console 窗口

在 Console 窗口中，若选择某条文本，则可以在下方看到更详细的说明，如图 1-139 所示。

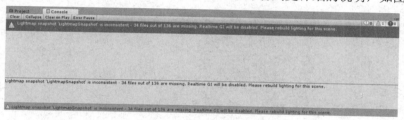

图 1-139　Console 窗口输出的信息

1.4　自定义窗口布局

Unity 的窗口布局是可以自定义的，开发者可以根据自己的使用习惯对窗口进行布局，也可以使用 Unity 内置的窗口布局功能对窗口布局进行调整。

1.4.1　使用 Unity 内置的窗口布局功能

工具栏的最右侧是 Layout 按钮，单击该按钮可以弹出一个浮动菜单，该菜单中包含 Unity 内置的窗口布局方式，如图 1-140 所示。

图 1-140　浮动菜单

5 种内置的窗口布局方式如图 1-141～图 1-145 所示。

图 1-141　2 + 3 窗口布局方式（选择 2 by 3 命令）

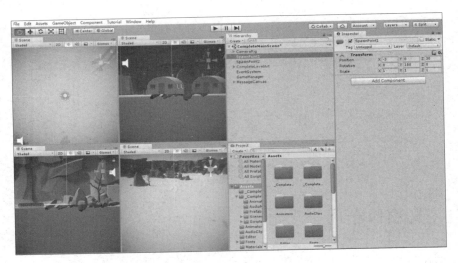

图 1-142　四视图窗口布局方式（选择 4 Split 命令）

图 1-143　默认窗口布局方式（选择 Default 命令）

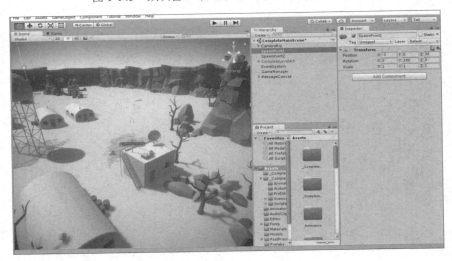

图 1-144　高屏窗口布局模式（选择 Tall 命令）

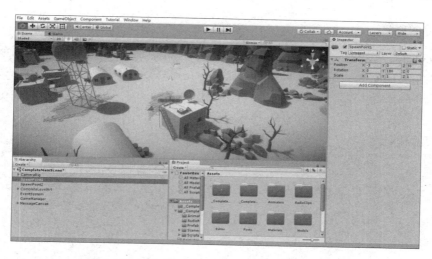

图 1-145　宽屏窗口布局模式（选择 Wide 命令）

1.4.2　自由设定窗口布局

Unity 中的窗口可以通过拖动的方式重新布局。

1．停靠窗口

如果想把 Project 窗口停靠在编辑器的左侧，那么可以使用鼠标左键选择该窗口的标题，并按住鼠标左键将其拖到编辑器的左侧。在拖动过程中，Project 窗口会以线框的方式显示，如图 1-146 所示。当 Project 窗口停靠到需要的地方时，松开鼠标左键，由此完成该窗口的布局操作，如图 1-147 所示。

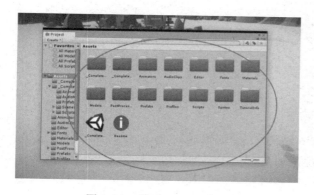

图 1-146　拖动 Project 窗口

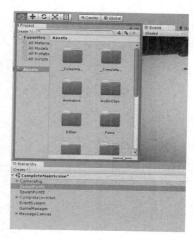

图 1-147　停靠 Project 窗口

2．浮动窗口

每个窗口可以浮动在编辑器中而不使用停靠的布局方式。还是以 Project 窗口为例，使用鼠标左键选择 Project 窗口的标题，并按住鼠标左键将其拖到需要的位置，松开鼠标左键，此时会形成一个浮动窗口，如图 1-148 所示。

3．内嵌窗口

在同一个窗口中可以内嵌其他窗口，如可以把 Hierarchy 窗口内嵌到 Project 窗口中。使用鼠标左键选择 Hierarchy 窗口的标题，按住鼠标左键将该窗口的标题拖到 Project 窗口的标题上，此时，Hierarchy 窗口和 Project 窗口会共用同一个区域。要切换这两个窗口，可以通过该区域上面的标题来完成，如图 1-149 所示。

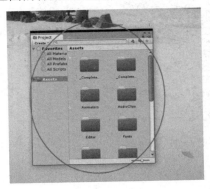

图 1-148　浮动 Project 窗口

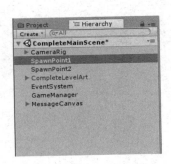

图 1-149　内嵌 Hierarchy 窗口

4．添加窗口

在每个窗口中，右上角有一个图标，单击该图标可以弹出窗口添加菜单，如图 1-150 所示。Maximize 命令用于最大化窗口，快捷键为空格键；Close Tab 命令用于关闭窗口；Add Tab 命令用于在该区域添加其他的窗口（也可以通过主菜单栏中的 Window 菜单来添加）。可以添加的窗口列表如图 1-151 所示。

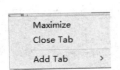

图 1-150　添加菜单

图 1-151　可以添加的窗口列表

1.5　本章小结

本章主要介绍了 Unity 的下载和安装，以及 Unity 编辑器的布局，同时简要介绍了如何自定义窗口布局，为读者学习后续内容奠定基础。

1.6　练习题

1．按照本章介绍的内容下载和安装 Unity，并注册账号，以备使用。如果有可能，请记录安装过程中遇到的困难，并与同学进行讨论。

2．简述 Project 窗口、Scene 窗口、Game 窗口、Inspector 窗口和 Console 窗口的作用。

3．（选做）尝试自定义一种窗口布局。

第 2 章

打地鼠

2.1 游戏简介

《打地鼠》游戏的原名为 *Whac-A-Mole*，其历史可追溯至 1976 年。当时，创意工程股份有限公司推出了一款备受追捧的街机游戏，该游戏凭借激动人心的体验赢得了大众的喜爱。这款游戏不仅能提升玩家大脑与身体的协调性，还能锻炼玩家的臂力与体能，所以逐渐成为广受欢迎的休闲活动。那么，这款游戏的玩法又是怎样的呢？在规定的时间内，玩家需要竭尽全力去敲打一只只从地洞中冒出头的地鼠，从而获得游戏的胜利。然而，如果在掌机（如 NDS 或智能手机等设备）上玩耍，就无须担心缺少锤子的问题，只需要使用 NDS 触控笔或手指轻轻一戳。图 2-1 所示为经典《打地鼠》游戏的画面。

2.2 游戏规则

这款游戏的游戏规则如下：在一定的时间内，地鼠随机出现在 9 个洞口中，玩家要在它出现的时候击中它，击中加分，反之地鼠会自动消失，时间耗尽时游戏结束。随着游戏时间的减少，地鼠出现的频率越来越高，存在的时间也越来越短，游戏难度逐步上升。

图 2-1 经典《打地鼠》游戏的画面

2.3 游戏开发核心思路

2.3.1 洞口位置的存储结构和计算

9 个洞口的位置和状态可用 3×3 的二维数组存储。此处使用一维数组存储洞口的位置信息及地鼠的出现状态。使用一维数组是为了后续使用随机数更加方便。

利用公式 k = i * n + j，可以把二维数组 a[m][n] 中的任意元素 a[i][j] 转化为一维数组的对应元素 b[k]。m 和 n 分别表示二维数组的行数和列数，i 为元素所在行，j 为元素所在列，同时 0 <= i < m 且 0 <= j < n。坐标值和对应值的例子如下：

```
[2][0]   [2][1]   [2][2]        →     [6]   [7]   [8]
[1][0]   [1][1]   [1][2]        →     [3]   [4]   [5]
[0][0]   [0][1]   [0][2]        →     [0]   [1]   [2]
```

2.3.2 地鼠出现频率的控制

可以利用改变随机数的取值范围来控制地鼠出现的频率和个数。例如，在区间[0,5)中取某个值的概率是1/5。在区间[0,10)中取该值的概率为1/10，前者比后者的概率要高一些。

当然，也可以利用延迟调用函数改变地鼠出现的频率和个数，这就需要使用 Unity 中的一个函数，即 InvokeRepeating()。该函数有 3 个参数，分别为 methodName、time 和 repeatRate。当 Unity 运行到 InvokeRepeating()函数时，首先延迟 time 的时间，接着以 repeatRate 的频率不断调用 methodName。

下面以伪代码为例展开介绍。start()函数从第 0 秒开始调用 canAppear()函数，且每 5 秒调用一次；canAppear()函数从被调用的第 0 秒开始调用地鼠生成函数，且每秒调用 1 次。也就是说第 0 秒开始，每秒生成 1 只地鼠；从第 5 秒开始，每秒生成 2 只地鼠……以此类推。伪代码如下：

```
public void start()
{
    InvokeRepeating("canAppear", 0,5);
}
public void canAppear()
{
    InvokeRepeating("地鼠生成函数", 0,1);
}
```

2.3.3 地鼠的随机生成和销毁

在生成对象时使用 Instantiate()函数，该函数需要传入 3 个参数，第 1 个是生成的对象，第 2 个是对象生成后放置的位置，第 3 个是对象生成后的初始朝向，返回任何继承自 Object 的对象。

在销毁对象时使用 Destroy()函数，该函数有 2 个参数，第 1 个是 Object，第 2 个是 t。当运行 Destroy()函数时，经过 t 的时间段后，销毁 Object。如果没有为 t 赋值，则直接销毁 Object。

在地鼠上设置鼠标监听及计时器。当地鼠出现后，如果在规定的时间内被击中，则地鼠立刻消失，反之到时间结束后自动消失。伪代码如下：

```
if（未被击中）
    Destroy(this.gameObject,3.0f);
else
    Destroy(this.gameobject);
```

2.3.4 游戏时长和分数

可以利用整型变量记录游戏时长和分数。每次在游戏开始后，玩家的分数会被清零，时长则设置为初始值。随着游戏的进行，时长参数的值减小，分数参数的值随着玩家击中地鼠数量的增加而增大。计时使用配合回调函数 Update()和静态变量 Time.deltaTime 来实现。

2.3.5 游戏流程图

游戏流程图如图 2-2 所示。

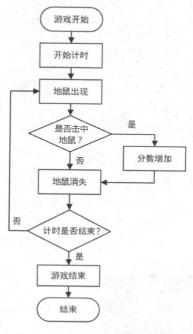

图 2-2 游戏流程图

2.4 游戏实现

2.4.1 资源的导入与 Sorting Layer

（1）新建工程。新建名称为 WhacAMole 的工程，选择 2D 选项，如图 2-3 所示。

图 2-3 新建工程

（2）导入素材包。创建不同的文件夹，将不同的资源放在不同的文件夹中。将下载好的图片资源保存到 Sprites 文件夹中，如图 2-4 所示。

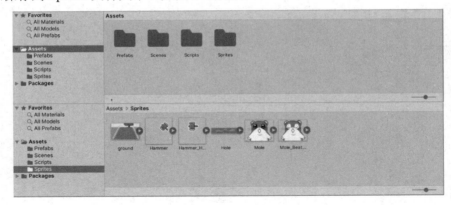

图 2-4　导入素材包

（3）调整背景和摄像机参数。在 Sprites 文件夹中选择 ground 图片，并将其拖入 Scene 窗口中。调整摄像机的视野范围，使 ground 图片占据整个视野。在 Hierarchy 窗口中单击 Main Camera 物体后，可以采用两种方式调整摄像机的视野范围：一种是直接在 Scene 窗口中调整白色线框的大小，如图 2-5 所示；另一种是在 Inspector 窗口中调整 Size 参数的值，如图 2-6 所示。

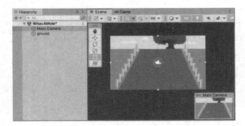 

图 2-5　使用白色线框调整摄像机的视野范围　　图 2-6　使用 Inspector 窗口中的 Size
　　　　　　　　　　　　　　　　　　　　　　　　　　参数调整摄像机的视野范围

（4）设置图片的 Sorting Layer。展开 Sorting Layers，单击图标➕，创建一个名为 Ground 的新图层。将 Ground 图层的 Sorting Layer 设置为 Ground，如图 2-7 所示。

需要注意的是，Unity 会根据 Order in Layer 的值从低到高绘制图像。值越小，说明该物体的渲染时间越早，显示在越低层，会被遮盖；值越大，说明该物体的渲染时间越晚，显示在越高层，可以覆盖低层的图片，如图 2-8 所示。

 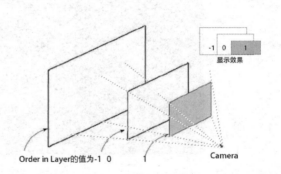

图 2-7　设置 Ground 图层的 Sorting Layer　　图 2-8　Order in Layer 的值与画面显示效果的关系示意图

（5）制作 Hole 和 Mole 的预制体。先将 Hole 图片拖到 Scene 窗口中，并为该图片设置 Sorting Layer，使 Hole 图片显示在 ground 图片的上方，再将 Hole 图片拖到 Prefabs 文件夹中，并删除 Scene 窗口中的 Hole 图片。采用同样的方式制作 Mole 的预制体。为不同资源设置 Sorting Layers，如图 2-9 所示。制作的预制体如图 2-10 所示。

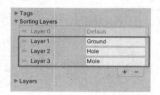

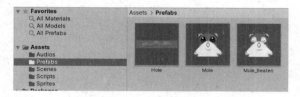

图 2-9　为不同资源设置 Sorting Layers　　　　图 2-10　制作的预制体

2.4.2　生成洞口

（1）创建脚本。在 Scripts 文件夹中，新建一个脚本并命名为 GameController，如图 2-11 所示。

图 2-11　创建脚本

（2）生成按 3×3 规则排列的洞口，并且保存洞口的信息。可以将 3×3 的洞口由二维数组映射为一维数组（使用一维数组只是为了方便后续使用一个随机数生成索引），使用 Instantiate()函数对洞口对象进行实例化。GameController 脚本中的代码如下：

```csharp
public class GameController : MonoBehaviour
{
    public struct Hole                              //描述洞口的结构体
    {
        public bool isAppear;                       //表示洞口是否出现了地鼠
        public int holeX;                           //表示洞口的 X 坐标
        public int holeY;                           //表示洞口的 Y 坐标
    }
    public Hole[] holes;                            //保存所有洞口的信息
    private float intervalPosX = 2, intervalPosY = 1;//每个洞口的横向间隔和纵向间隔
    public GameObject holeObj;                      //要实例化的洞口预制体

    //Start is called before the first frame update
    void Start()
    {
```

```
        InitMap();                                    //初始化洞口
    }
    //初始化场景洞口
    private void InitMap()
    {
        Vector2 originalPos = new Vector2(-2,-2);     //洞口的初始坐标
        holes = new Hole[9];                          //分配存储洞口信息的内存
        //初始化每个洞口的信息并创建洞口对象
        for (int i = 0; i < 3; i++)
        {
            for (int j = 0; j < 3; j++)
            {
                holes[i * 3 + j] = new Hole();
                //计算每个洞口的X坐标
                holes[i * 3 + j].holeX = (int)(originalPos.x + j * intervalPosX);
                //计算每个洞口的Y坐标
                holes[i*3+j].holeY = (int)(originalPos.y + i*intervalPosY);
                holes[i*3+j].isAppear = false;        //表示当前洞口没有地鼠
                Instantiate(holeObj,newVector3(holes[i*3+j].holeX,holes[i*3+j].holeY,0),
                    Quaternion.identity);             //实例化洞口对象
            }
        }
    }
}
```

图 2-12　在 ground 对象上挂载脚本并赋值

（3）先将该脚本挂载到 ground 对象上，再将 Hole 预制体赋值给脚本中的 Hole Obj，如图 2-12 所示。

2.4.3　地鼠的生成

《打地鼠》游戏的核心功能是地鼠的出现和消失，本节介绍地鼠的生成和被击中后消失的流程。

（1）实现地鼠的生成。需要使地鼠随机生成在洞口，并控制地鼠生成的频率。可以先使用随机数决定地鼠生成于哪个洞口，再使用 InvokeRepeating()函数控制地鼠的生成。在脚本修改完后，为 moleObj 赋值。GameController 脚本中的代码如下（加粗字体为在原有代码的基础上新增的内容，后续沿用本规定）：

```
public class GameController : MonoBehaviour
{
    public struct Hole                               //描述洞口的结构体
    {
        public bool isAppear;                        //表示洞口是否出现了地鼠
        public int holeX;                            //表示洞口的X坐标
        public int holeY;                            //表示洞口的Y坐标

        public GameObject mole;                      //该洞口出现的地鼠
    }
    public Hole[] holes;                             //保存所有洞口的信息
    private float intervalPosX = 2, intervalPosY = 1;//每个洞口的横向间隔和纵向间隔
    public GameObject holeObj;                       //要实例化的洞口预制体
```

```csharp
    public GameObject moleObj;                          //要实例化的地鼠预制体

//Start is called before the first frame update
void Start()
{
    InitMap();                                          //初始化洞口

    InvokeRepeating("MoleAppear", 0f, 0.5f);            //测试地鼠出现的频率
}

//Update is called once per frame
void Update()
{
    CleanHoleState();                                   //检测并清除洞口状态
}


//初始化场景洞口
private void InitMap()
{
    Vector2 originalPos = new Vector2(-2,-2);           //洞口的初始坐标
    holes = new Hole[9];                                //分配存储洞口信息的内存
    //初始化每个洞口的信息并创建洞口对象
    for (int i = 0; i < 3 ; i++)
    {
        for (int j = 0; j < 3; j++)
        {
            holes[i * 3 + j] = new Hole();
            //计算每个洞口的X坐标
            holes[i * 3 + j].holeX = (int)(originalPos.x + j * intervalPosX);
            //计算每个洞口的Y坐标
            holes[i*3+j].holeY = (int)(originalPos.y + i * intervalPosY);
            holes[i*3+j].isAppear = false;              //表示当前洞口没有地鼠
            Instantiate(holeObj,new Vector3(holes[i*3+j].holeX,holes[i*3+j].holeY,0),
                    Quaternion.identity);               //实例化洞口对象
        }
    }
}

private void MoleAppear()
{   //控制地鼠的出现
    int id = UnityEngine.Random.Range(0, 9);            //获得随机数
    while (holes[id].isAppear == true)
    {   //判断当前洞口是否已经有地鼠了，为了防止死循环，需要在Mole类中使用Destroy()函数
        id = UnityEngine.Random.Range(0, 9);            //如果有，则重新获得随机数
    }
    //在对应ID的洞口实例化地鼠对象
    holes[id].mole = Instantiate(moleObj, new Vector3(holes[id].holeX,
                                 holes[id].holeY, 0),
                                 Quaternion.identity);
    //在对应ID的洞口中将isAppear设置为true，这样设置是为了判断该洞口中是否有地鼠
    holes[id].isAppear = true;
```

```
        Debug.Log("MoleAppear");
    }

    private void CleanHoleState()
    {   //清除洞口地鼠信息
        for (int i = 0; i < 3; i++)
        {
            for (int j = 0; j < 3; j++)
            {
                if (holes[i * 3 + j].mole == null)
                {
                    holes[i * 3 + j].isAppear = false;
                }
            }
        }
    }
}
```

（2）实现地鼠的消失。新建 Mole 脚本，调用 Destroy()函数，使地鼠在生成后按照一定的时间消失。选择 Mole 预制体，单击 Add Component 按钮，找到 Mole 脚本，单击 Mole 脚本将其挂载到 Mole 预制体上，如图 2-13 所示。Mole 脚本中的代码如下：

```
public class Mole : MonoBehaviour
{
    //Start is called before the first frame update
    void Start()
    {
        Destroy(gameObject, 3f);
    }

    //Update is called once per frame
    void Update()
    {

    }
}
```

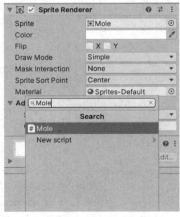

图 2-13　将 Mole 脚本挂载到 Mole 预制体上

2.4.4 打击地鼠

（1）为地鼠设定碰撞盒。为了实现打击地鼠功能，需要使用 OnMouseDown() 函数，该函数可以用来判断是否单击到物体的碰撞盒。选择 Mole 预制体，单击 Add Component 按钮，在搜索框中输入 collider，选择 Polygon Collider 2D 选项，如图 2-14 所示，为地鼠添加碰撞盒。为地鼠添加碰撞盒后的效果如图 2-15 所示。

图 2-14　选择 Polygon Collider 2D 选项　　　图 2-15　为地鼠添加碰撞盒后的效果

（2）消除地鼠。使用 OnMouseDown() 函数判断是否击中地鼠，如果击中地鼠，则使用 Destroy() 函数将其销毁。Mole 脚本中的代码如下：

```
public class Mole : MonoBehaviour
{
    public GameObject beatenMole;              //用于显示被打击地鼠的对象
    public int id;                             //对应洞口的 ID
    public GameController gameController;      //获得 gameController 的对象

    //Start is called before the first frame update
    void Start()
    {
        //获得 gameController 的对象
        gameController = GameObject.FindObjectOfType<GameController>();

        Destroy(gameObject, 3f);
    }

    void OnMouseDown()                         //打击地鼠的功能
    {
        //产生被击中地鼠的对象
        gameController.holes[id].mole = Instantiate(beatenMole,gameObject.transform
.position, Quaternion.identity);
        Destroy(gameObject);                   //销毁正常的地鼠
        Debug.Log("OnMouseDown");              //测试代码
    }
}
```

（3）打击地鼠后的效果。创建新脚本 BeatenMole，当地鼠被击中后，在同样的位置生成 BeatenMole 预制体，并控制其在一定时间后消失。将 BeatenMole 脚本挂载在 BeatenMole 预制体上。BeatenMole 脚本中的代码如下：

```
public class BeatenMole : MonoBehaviour
{
    //Start is called before the first frame update
```

```
    void Start()
    {
        Destroy(gameObject, 1f);
    }
}
```

2.4.5 计时功能

（1）编辑脚本。创建一个名称为 Timer 的新脚本，需要在游戏界面出现一个倒计时，以配合回调函数 Update()及静态变量 Time.deltaTime 递减时间，并使用 TextMeshPro 显示时间。Timer 脚本中的代码如下：

```
using UnityEngine.UI;                                       //导入 UI 包
using TMPro;                                                //导入 TMPro 包
public class Timer : MonoBehaviour
{
    public TextMeshProUGUI timerText;                       //显示计时的界面
    public float time = 30.0f;                              //游戏时间
    private bool canCountDown = false;                      //判断是否可以计时
    // Start is called before the first frame update
    void Start()
    {

    }

    // Update is called once per frame
    void Update()
    {
        if (canCountDown == true)
        {   //判断是否可以计时
            time -= Time.deltaTime;                         //开始计时
            timerText.text = "Time: " + time.ToString("F1");//在 UI 上显示计时信息
            //Debug.Log(time);
        }
    }

    public void CountDown(bool countDown)
    {   //用于设置是否开始倒计时
        this.canCountDown = countDown;
    }
}
```

修改后 GameController 脚本中的代码如下：

```
public class GameController : MonoBehaviour
{
    public Timer timer;                                     //保存计时器

    //Start is called before the first frame update
    void Start()
    {
        InitMap();                                          //初始化洞口
        InvokeRepeating("MoleAppear", 0f, 0.5f);            //测试地鼠出现的频率
```

```
        timer.CountDown(true);                              //开始计时
    }
}
```

（2）创建 Timer 对象。在场景中创建一个名称为 Timer 的新对象，并将 Timer 脚本挂载到这个对象上，如图 2-16 所示，同时为 GameController 脚本中的 Timer 赋值。

（3）创建计时 TextMeshPro。创建一个 TextMeshPro，导入 TMP Essentials，先将新创建的游戏对象命名为 TimeCountDownText，并将 TextMeshPro 的内容修改为 Time: 0，然后把 TimeCountDownText 赋值给 Timer 脚本中的 Timer Text 参数，最后将 TimeCountDownText 放在合适的位置。具体流程如图 2-17～图 2-21 所示。

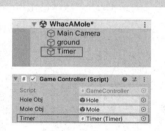

图 2-16　创建 Timer 对象并挂载 Timer 脚本

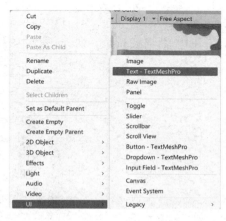

图 2-17　创建 TextMeshPro

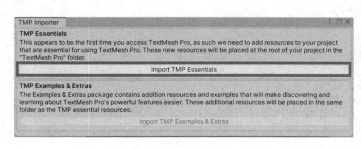

图 2-18　导入 TMP Essentials

图 2-19　设置 TextMeshPro

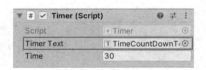

图 2-20　把 TimeCountDownText 赋值给 Timer 脚本中的 Timer Text 参数

图 2-21　在游戏场景中 TimeCountDownText 的显示效果

2.4.6 计分功能

(1) 使用 TextMeshPro 创建计分 Text 对象。创建一个新的 TextMeshPro，并将其命名为 TimerText；将新建的 TextMeshPro 的内容修改为 Score: 0，如图 2-22 所示。

图 2-22 使用 TextMeshPro 创建计分 Text 对象

(2) 编辑脚本。需要实现在游戏中显示分数，且打中地鼠后分数增加。可以使用静态变量计分，并使用 GameObject.Find()函数寻找对象。Mole 脚本中的代码如下：

```
using UnityEngine.UI;
using TMPro;

public class Mole : MonoBehaviour
{
    public GameObject beatenMole;         //用于显示被打击地鼠的对象
    public int id;                        //对应洞口的 ID
    public GameController gameController; //获得 gameController 的对象

    public TextMeshProUGUI scoreText;     //显示分数的 UI
    public static int score = 0;          //记录分数的变量

    // Start is called before the first frame update
    void Start()
    {
        //获得显示分数的 UI
        scoreText = GameObject.Find("ScoreText").GetComponent<TextMeshProUGUI>();

        //获得 gameController 的对象
        gameController = GameObject.FindObjectOfType<GameController>();
        Destroy(gameObject, 3f);
    }

    void OnMouseDown()                    //鼠标按键按下时监听
    {
        score++;                          //加分
        scoreText.text = "Score: " + score; //显示分数
        //产生被击中地鼠的对象
        gameController.holes[id].mole = Instantiate(beatenMole,
                                        gameObject.transform.
                                        position,Quaternion.identity);
        Destroy(gameObject);              //销毁正常的地鼠
        Debug.Log("OnMouseDown");         //测试代码
    }
}
```

2.4.7 游戏结束

修改脚本，添加游戏结束的判断。需要实现在计时器归零时停止生成地鼠。可以使用 CancelInvoke()函数停止调用 MoleAppear()函数。Timer 脚本中的代码如下：

```
public class Timer : MonoBehaviour
{
    public void CountDown(bool countDown)
    {   //用于设置是否开始倒计时
        this.canCountDown = countDown;
        if (canCountDown == false)
        {   //计时器归零
            time = 0;timerText.text = "Game Over!!!";
        }
    }
}
```

GameController 脚本中的代码如下：

```
public class GameController : MonoBehaviour
{
    void Update()
    {
        CleanHoleState();               //检测并清除洞口状态

        if (timer.time < 0)
        {   //判断是否结束倒计时
            GameOver();                 //游戏结束提示
        }
    }

    private void GameOver()             //游戏结束的操作
    {
        timer.CountDown(false);         //时间归零
        CancelInvoke();                 //停止生成地鼠
    }
}
```

游戏结束画面如图 2-23 所示。

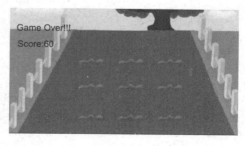

图 2-23　游戏结束画面

2.4.8　修改地鼠出现频率

当游戏时长小于 15 秒时，需要提高地鼠出现频率，这就需要设置控制地鼠出现频率的变量。GameController 脚本中的代码如下：

```csharp
public class GameController : MonoBehaviour
{
    public float appearFrequency = 0.5f;
    private bool canIncreaseMole = true;                    //判断是否修改地鼠出现频率

    //Start is called before the first frame update
    void Start()
    {
        InitMap();                                          //初始化洞口
        MoleAppearFrequency(appearFrequency);               //控制地鼠出现频率
        //InvokeRepeating("MoleAppear", 0f, 0.5f);          //测试地鼠出现频率

        timer.CountDown(true);                              //开始计时
    }

    //Update is called once per frame
    void Update()
    {
        CleanHoleState();                                   //检测并清除洞口状态

        if (timer.time < 15 && canIncreaseMole == true)
        {   //修改地鼠出现频率
            appearFrequency -= 0.5f;
            MoleAppearFrequency(appearFrequency);canIncreaseMole = false;   //判断是否修改地鼠出现频率
        }

        if (timer.time < 0)
        {                                                   //判断是否结束倒计时
            GameOver();                                     //游戏结束提示
        }
    }

    private void MoleAppearFrequency(float appearFrequency)
    {   //控制地鼠出现频率,加入修改地鼠出现频率的参数
        CancelInvoke();                                     //停止生成地鼠
        InvokeRepeating("MoleAppear",0,appearFrequency);    //重新开始生成地鼠
    }
}
```

2.4.9 修改图标

（1）用代码修改图标。创建一个新的脚本并命名为 ChangeCursor。可以通过修改 Cursor.visible 隐藏箭头鼠标，同时使用 Input.mousePosition 获取鼠标指针的屏幕坐标。ChangeCursor 脚本中的代码如下：

```csharp
using UnityEngine.UI;

public class ChangeCursor : MonoBehaviour
{
    public Sprite normalCursor;                             //未击打的锤子图标
    public Sprite hitCursor;                                //击打的锤子图标
    public Image hammerImage;                               //显示锤子图标
```

```csharp
//Start is called before the first frame update
void Start()
{
    Cursor.visible = false;                    //隐藏箭头鼠标
}

//Update is called once per frame
void Update()
{
    if (Input.GetMouseButton(0))
    {   //按下鼠标左键时
        hammerImage.sprite = hitCursor;        //更换成击打的图标
    }
    else
    {
        hammerImage.sprite = normalCursor;     //更换成未击打的图标
    }
    //锤子图标在屏幕上的位置由鼠标指针在屏幕上的位置确定
    hammerImage.rectTransform.position = Input.mousePosition;
}
```

（2）改变鼠标指针的图标。创建一个 Image 对象并将其命名为 HammerImage，同时将 Hammer 图片赋值给该对象。将 ChangeCursor 脚本挂载到 HammerImage 对象上，同时为所需要的参数赋值，如图 2-24 所示。最终的游戏效果如图 2-25 所示。

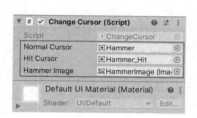

图 2-24　挂载 ChangeCursor 脚本

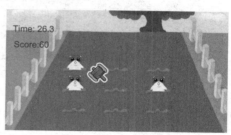

图 2-25　最终的游戏效果

2.4.10　添加音效

添加 Audio Source 组件。新建一个文件夹 Audios 用来保存音效，并将音频导入该文件夹中。打开预制体 Mole，单击 Add Component 按钮，搜索 Audio Source，将音频 appear 赋值给 Audio Clip，确保已勾选 Play On Awake 复选框，使用同样的方式为 Mole Beaten 添加组件，如图 2-26 所示。

图 2-26　添加 Audio Source 组件

2.4.11 重新开始游戏

（1）创建 RestartButton。首先在 Canvas 下创建一个 Button 并命名为 RestartButton，然后将其放置在合适的位置；将 Button 的名称修改为 Restart，如图 2-27 所示。

（2）实现重新开始。创建一个新的脚本并命名为 Restart。在单击"重新开始"按钮后，使用 SceneManager 重新载入场景。Restart 脚本中的代码如下：

```
using UnityEngine.SceneManagement;

public class Restart : MonoBehaviour
{
    public void RestartGame()
    {   //重新开始游戏
        SceneManager.LoadScene("WhacAMole");   //输入场景名称
    }
}
```

（3）添加按钮事件。将场景名称修改为 WhacAMole。将 Restart 脚本挂载到 RestartButton 上，同时设置按钮事件，在单击"重新开始"按钮后，调用 Restart 脚本中的 RestartGame() 函数，如图 2-28 所示。

图 2-27　创建 RestartButton

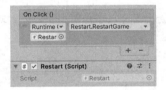

图 2-28　添加按钮事件

2.4.12 导出游戏

（1）导出游戏。先在 Build Settings 中添加当前场景，再单击 Build 按钮，选择导出的目录，如图 2-29 所示。

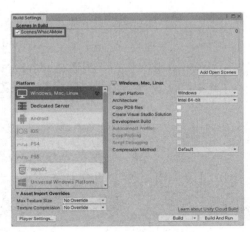

图 2-29　导出游戏

（2）添加退出游戏功能。新建一个脚本并命名为 Quit，需要实现按 Esc 键退出程序，即使用 Application.Quit()函数退出程序。Quit 脚本中的代码如下：

```
public class Quit : MonoBehaviour
{
    //Update is called once per frame
    void Update()
    {
        if (Input.GetKeyDown(KeyCode.Escape))
        {
            Application.Quit();
        }
    }
}
```

2.5 本章小结

通过学习本章内容，读者可以体验利用 Unity 设计与制作简单游戏《打地鼠》的完整流程，以及了解游戏的设计与制作。本章主要介绍了资源导入方式、计时与计分功能、添加音效、开始与结束游戏等内容。《打地鼠》游戏的重点在于洞口的生成、地鼠的生成与消失、打击判断逻辑等。

2.6 练习题

1. 本章在介绍图片资源时介绍了 Sorting Layer 的概念，现在有两个图层 A 和 B，这两个图层在排序层列表中分别排在第 2 位和第 3 位。假设处在图层 A 和图层 B 中的具有相同大小、形状及朝向的 2 张图片在同一场景下完全重合（坐标相同），那么此时显示的是哪个图层中的图片呢？

2. 假设第 1 只地鼠从第 3 秒开始出现，且前后 2 只地鼠出现的间隔时间为 3 秒，停留时间为 7 秒，洞口为 12 个，那么同一时刻出现的地鼠最多为多少只？

3. 在一般情况下，一维数组与二维数组的转换是以行优先的方式进行映射的，即在计算二维数组中的某个元素在一维数组中的索引时确定其前面有多少行。现换一种思考方式，以列优先的方式进行映射，那么一个 m 行 n 列的二维数组中的第 r 行第 c 列的元素在一维数组中的索引是多少？

4. 在《打地鼠》游戏运行时，若采用本章中的代码，则同时出现的地鼠数目一定要小于洞口数目，否则会发生什么现象？请解释原因。

5.（选做）若洞口的位置在场景中的位置分布毫无规律可言，则在用程序实现时应采取何种策略来表达此种情况下的洞口？请从数据结构的角度进行阐述。

第 3 章

记忆翻牌

3.1 游戏简介

《记忆翻牌》是一款让人爱不释手的经典游戏。该游戏不但有趣,而且可以帮助玩家训练记忆力。不管玩家是大人还是小孩,都可以轻松上手。玩家只需要找到两张可以配对的卡片,便可以得分。在成功找到所有配对的卡牌后,游戏便圆满结束。图 3-1 所示为《记忆翻牌》游戏的画面。

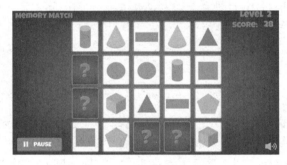

图 3-1 《记忆翻牌》游戏的画面

3.2 游戏规则

游戏界面内共有 12 张卡牌,6 种图案,两两成对。玩家每次可以翻开两张卡牌,若配对,则两张卡牌将始终处于正面,否则,再次翻转为背面。在所有卡牌配对成功后,计时停止,游戏结束。游戏记录步数,步数越少成绩越好。当不同的两张卡牌被翻出时,需要等待一段时间后才能继续翻转卡牌。

3.3 游戏开发核心思路

3.3.1 生成卡牌池

《记忆翻牌》游戏的卡牌池一般由 3 行 4 列的 12 张卡牌组成,可以使用自动布局配置卡牌列表，随机排列两张相同的卡牌。通过二维数组为每张卡牌进行编号，随机赋予卡牌的性质，以便后续追踪每张卡牌的状态：以每张卡牌的中心为锚点，以左下角卡牌为起始点。左下角卡牌的坐标为 originalX=-3，originalY=0；相邻两张卡牌横向间隔设置为 offsetX=2，纵向间隔设置为 offsetY=2.5；每张卡牌相对于左下角卡牌的位置为(x,y) = (offsetX * j + originalX, offsetY * i + originalY)，其中，i 和 j 分别表示第 i 行和第 j 列。卡牌位置如图 3-2 所示。

3.3.2 洗牌

在《记忆翻牌》游戏中，共有 12 张卡牌，6 种不同的牌面，每种牌面 2 张。可以采用多种方式进行洗牌，下面介绍一种简单的洗牌方法。先创建一个共有 6 个元素的一维数组 images，该数组用于保存 6 种不同的牌面。再创建一个共有 12 个元素的一维数组 numbers，并将该数组初始化为{ 0, 0, 1, 1, 2, 2, 3, 3, 4, 4, 5, 5 }，每个元素的值代表 images 数组对应的索引

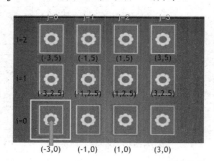

图 3-2　卡片位置

值。对 numbers 数组进行洗牌，把元素值打乱。按照场景卡牌的布局，在每个对应的位置生成卡牌的牌面对象。根据 i 和 j，先从 numbers 数组中取出对应 images 数组的索引值，再按照该索引值找到每个牌面的图案。

在本游戏中，洗牌方式如下：遍历整个 numbers 数组，获得当前索引 i 的元素值；在当前元素的索引和最后一个元素的索引之间取随机数 r；对调索引 i 和索引 r 的元素值；直到索引 i 为最后一个元素的索引。图 3-3 所示为洗牌前后的效果，图 3-4 所示为使用随机数算法进行洗牌的实现思路。

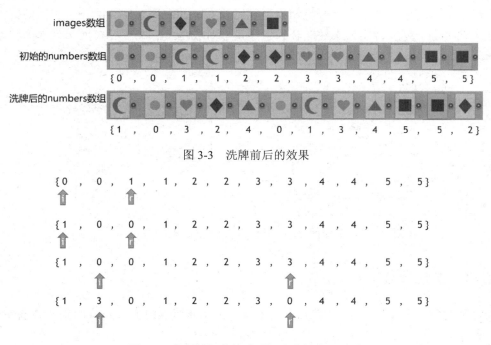

图 3-3　洗牌前后的效果

图 3-4　使用随机数算法进行洗牌的实现思路

通过 numbers 数组获得 images 数组中的卡片类型，即 images[numbers[i*gridCols + j]]，如图 3-5 所示，第 2 行第 1 列所对应的 numbers 数组的索引值为 $2×4+1=9$，即该卡牌对应 numbers 数组中的元素为 numbers[9] = 5，这样即可获得 images 数组所对应的第 5 张卡牌，即 images[5] = 方块卡牌，最后把该图片对象赋值给对应位置的卡牌对象即可。

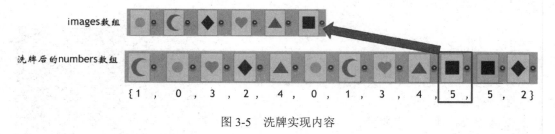

图 3-5　洗牌实现内容

注意：numbers 数组只保存了 images 数组对应的索引号，而 images 数组才保存了实际的图片素材。

3.3.3　卡牌状态

《记忆翻牌》游戏的核心内容是卡牌配对。这里为每张卡牌定义 3 种状态，分别为未被翻开状态、翻开状态和配对成功状态。可以建立一个卡牌类，为每张卡牌定义初始状态，即未被翻开状态。

在场景设置中，一张卡牌其实是由两张图重叠组成的，在上面的是卡牌的背面，在下面的则是卡牌的图案。当玩家单击卡牌时，上面的图（卡牌的背面）取消显示，让下面的卡牌的图案显示出来。

当玩家单击两张卡牌以后，卡牌切换至翻开状态，此时需要判断两张卡牌的图案是否相同，以此来决定卡牌应切换为未被翻开状态还是配对成功状态。图 3-6 所示为卡牌状态切换示意图。

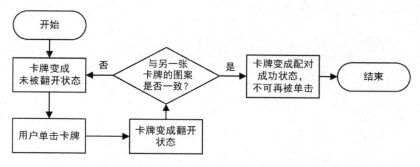

图 3-6　卡牌状态切换示意图

3.3.4　游戏计分

为每种类型设置不同的 ID，当有两张卡牌处于翻开状态时，判断这两张卡牌的 ID 是否相等，若相等，则表示配对成功，否则配对失败。

伪代码如下：

```
if (firstRevealed.id == secondRevealed.id)
{
        //配对成功的操作,如加分等
}else
{
        //配对失败的操作,如重新把卡牌翻转为背面等
}
```

3.3.5 游戏流程图

游戏流程图如图 3-7 所示。

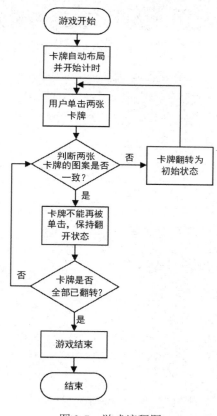

图 3-7 游戏流程图

3.4 游戏实现

3.4.1 游戏资源的导入与场景搭建

(1) 创建工程。新建一个名称为 MemoryCard 的 2D 工程,选择 2D 选项,如图 3-8 所示。
(2) 导入素材包并搭建场景。导入资源包 Memory,并确保贴图的类型都为 Sprite。将 table_top 放到 Hierarchy 窗口中,并设置它的位置和摄像机的位置一致,如图 3-9 所示。

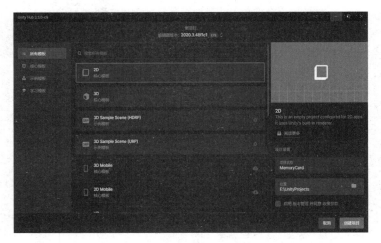

图 3-8　创建工程

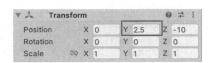

图 3-9　设置 table_top 与摄像机的位置

3.4.2　卡牌的位置排列

（1）调整摄像机。创建一个脚本并命名为 SceneManager。可以使用一维数组来保存卡牌数据，以每张卡牌的中心为锚点，以左下角的卡牌为起始点。SceneManager 脚本中的代码如下：

```csharp
public class SceneManager : MonoBehaviour
{
    public int gridRows =3;              //行
    public int gridCols = 4;             //列

    public float offsetX = 2f;           //卡牌 X 轴方向上的间距
    public float offsetY = 2.5f;         //卡牌 Y 轴方向上的间距

    public float originalX = -3f;        //起始卡牌的 X 坐标
    public float originalY = 0;          //起始卡牌的 Y 坐标

    public GameObject card;              //存储卡牌对象

    //Start is called before the first frame update
    void Start()
    {
        ArrangeCards();                  //排列卡牌
    }

    void ArrangeCards()
    {
        for(int i=0;i<gridRows;i++)
        {
            for(int j=0;j<gridCols;j++)
            {
```

```
                Gameobject tempCard;
                tempCard = Instantiate(card) as MemoryCard;
                //创建临时对象tempCard用来保存动态生成的卡牌
                float posX = (offsetX * j) + originalX;
                float posY = (offsetY * i) + originalY;
                //计算生成的卡牌在X轴和Y轴上的位置
                tempCard.transform.position = new Vector3(posX,posY,0);
                //把坐标赋值给tempCard对象
            }
        }
    }
}
```

（2）在 Unity 中新建一个空的游戏对象并命名为 SceneManager，把 SceneManager 脚本赋值给 SceneManager 对象。将 Sprite 卡牌存储为预制体并赋值给 SceneManager 脚本。图 3-10 所示为游戏的运行效果。

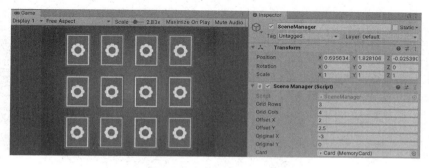

图 3-10　游戏的运行效果

3.4.3　洗牌功能的实现

在 SceneManager 脚本中，先创建一个共有 6 个元素的一维数组 images 来保存 6 种卡牌的图案，再创建共有 6×2 个元素的一维数组 numbers 并初始化为{0, 0, 1, 1, 2, 2, 3, 3, 4, 4, 5, 5}，每个元素值代表 images 数组对应的索引值，把 numbers 数组中元素值打乱。SceneManager 脚本中的代码如下：

```
public class SceneManager : MonoBehaviour
{
    ……
    public Sprite[] images;                        //保存卡牌的图案
    int[] numbers = {0,0,1,1,2,2,3,3,4,4,5,5};     //保存索引值
    ……
    void Start()
    {
        numbers = ShuffleCards(numbers);           //洗牌
        ArrangeCards();                            //排列卡牌
    }
    ……

    //洗牌
    int[] ShuffleCards(int[] numbers)
    {
        //使用tempNums复制一个numbers数组
        int[] tempNums = numbers.Clone() as int[];
```

```
        for(int i=0;i<tempNums.Length;i++)
        {
            int temp = tempNums[i];
            //获取tempNums 数组长度范围内的随机数
            int r = Random.Range(i,tempNums.Length);
            tempNums[i] = tempNums[r];
            tempNums[r] = temp;                          //交互索引值
        }
        //查看是否已经洗牌
        for(int j=0;j<tempNums.Length;j++)
        {
            Debug.Log(tempNums[j]);
        }
        return tempNums;
    }
}
```

3.4.4 显示卡牌的背面与卡牌的图案

（1）拖入图片。在 Inspector 窗口中找到 SceneManager 脚本，将 Sprites 文件夹中卡牌的图案拖到 images 数组中，如图 3-11 所示。

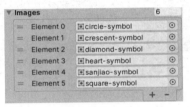

图 3-11 拖动卡牌的图案

（2）可以通过 numbers 数组获得 images 数组中卡牌的图案，即 images[numbers[i*gridCols + j]]。卡牌的背面显示可以新建一个类来管理卡牌的图案与背面。新建一个脚本并命名为 MemoryCard。修改 card 的类型，并在 Unity 中重新为其赋值。MemoryCard 脚本中的代码如下：

```
public class MemoryCard : MonoBehaviour
{
    public Sprite cardbackImage;          //管理卡牌的背面
    private Sprite cardImage;             //管理卡牌的图案

    //Start is called before the first frame update
    void Start()
    {

    }
    //Update is called once per frame
    void Update()
    {

    }

    //初始化卡牌的图案
    public void InitCardImage(Sprite image)
```

```
    {
        cardImage = image;              //获取当前卡牌的图案
    }
    //当卡牌翻到背面时
    public void Unreveal(){
        GetComponent<SpriteRenderer>().sprite = cardbackImage;
        cardState = CardState.Unreveal;
    }
    //当卡牌翻到正面时
    public void Reveal()
    {
        GetComponent<SpriteRenderer>().sprite = cardImage;
        cardState = CardState.Reveal;
    }
}
```

SceneManager 脚本中的代码如下：

```
public class SceneManager : MonoBehaviour
{
    public MemoryCard card;   //将类型修改为 MemoryCard
    ……

    void ArrangeCards()
    {
        for(int i=0;i<gridRows;i++)
        {
            for(int j=0;j<gridCols;j++)
            {
                MemoryCard tempCard;
                tempCard = Instantiate(card) as MemoryCard;
                //将类型修改为 MemoryCard
                tempCard.GetComponent<MemoryCard>().InitCardImage(images[numbers[i*gridCols+j]]);
                //从 numbers 数组中取出对应的 image 数组对应的索引值，按照索引值找到每张卡牌的图案
                tempCard.GetComponent<MemoryCard>().Unreveal();
                //将卡牌翻转到背面
                //tempCard.GetComponent<MemoryCard>().Reveal();
                //将卡牌翻转，检测是否已为卡牌赋值
                float posX = (65rigin * j ) + 65riginal;
                float posY = (65rigin * I ) +65riginallyY;
                tempCard.transform.position = new Vector3(posX,posY,0);
            }
        }
        ……
    }
}
```

（3）设置预制体。将卡牌的预制体命名为 Card，并且将 MemoryCard 脚本挂载到该预制体上，将卡牌的图案赋值给 MemoryCard 脚本。将 Card 预制体的 Order in Layer 设置为 1，如图 3-12 所示。

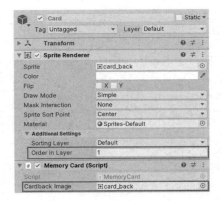

图 3-12　设置预制体

3.4.5　设置卡牌状态

（1）卡牌共有 3 种状态，分别为未被翻开状态、翻开状态及配对成功状态，可以使用枚举类型设置这 3 种状态。MemoryCard 脚本中的代码如下：

```
public class MemoryCard : MonoBehaviour
{
    public Sprite cardbackImage;
    private Sprite cardImage;

    public enum CardState{Reveal,Unreveal,Matched};       //枚举卡牌的 3 种状态
    CardState cardState = CardState.Unreveal;             //卡牌的默认状态是未被翻开

    //Start is called before the first frame update
    void Start()
    {

    }
    //使用回调函数 OnMouseDown()判断卡牌的当前状态
    public void OnMouseDown()
    {
        //若卡牌当前为未被翻开状态且不是配对成功状态，则可以翻开卡牌
        if (cardState == CardState.Unreveal && cardState != CardState.Matched &&
            sceneManager.CardReveal())
        {
            Reveal();                                     //调用 Reveal()函数来翻开卡牌
        }
    }

    public void InitCardImage(int id,Sprite image)
    {
        cardImage = image;
    }

    public void Unreveal()
    {
        GetComponent<SpriteRenderer>().sprite = cardbackImage;
        cardState = CardState.Unreveal;                   //将卡牌状态设置为未被翻开
```

```
    }
    public void Reveal(){
        GetComponent<SpriteRenderer>().sprite = cardImage;
        cardState = CardState.Reveal;                    //将卡牌状态设置为翻开
    }
    //将卡牌设置为配对成功状态
    public void Matched(){
        cardState = CardState.Matched;                   //将卡牌状态设置为配对成功
    }
}
```

（2）为了调用回调函数，需要为 Card 预制体添加 Box Collider 2D 碰撞盒，如图 3-13 所示。

图 3-13　为 Card 预制体添加 Box Collider 2D 碰撞盒

3.4.6　卡牌图案的配对

如果翻开的两张卡牌的图案相同，则配对成功，否则配对失败。应为每种卡牌的图案设置不同的 ID，当翻开两张卡牌后，若两张卡牌的 ID 相等，则表示配对成功。MemoryCard 脚本中的代码如下：

```
public class MemoryCard : MonoBehaviour
{
    ……
    private int ID;                              //用于保存每张卡牌的 ID
    private SceneManager sceneManager;
    //用于调用 SceneManager 脚本中的 CardReveal() 函数
    //Start is called before the first frame update
    void Start()
    {
        sceneManager = GameObject.FindObjectOfType<SceneManager>();
                                                 //获取 SceneManager 对象
    }
    public void OnMouseDown()
    {
        //只有sceneManager.CardReveal()为真，才可以继续翻开卡牌
        if(cardState == CardState.Unreveal && cardState != CardState.Matched &&
                    sceneManager.CardReveal())
        {
            Reveal();
            sceneManager.CardReveal(this);
```

```csharp
                //将当前选中的卡牌赋值给 SceneManager 脚本中的 CardReveal() 函数
            }
    }
    ……
    public void InitCardImage(int id,Sprite image)
    {
        ID = id;
        cardImage = image;
        Debug.Log("ID: = " + ID);
    }

    public int getID()    //获取卡牌的 ID
    {
        return ID;
    }
}
```

SceneManager 脚本中的代码如下:

```csharp
public class SceneManager : MonoBehaviour
{
    ……
    private MemoryCard firstRevealedCard;          //翻开第 1 张卡牌
    private MemoryCard secondRevealedCard;         //翻开第 2 张卡牌

    void ArrangeCards()
    {
        for(int i=0;i<gridRows;i++)
        {
            for(int j=0;j<gridCols;j++)
            {
                MemoryCard tempCard;
                tempCard = Instantiate(card) as MemoryCard;
                tempCard.GetComponent<MemoryCard>().InitCardImage(numbers[
                            i*gridCols+j],images[numbers[i*gridCols+j]]);
                //i*gridCols+j 对应相应的卡牌的 ID
                tempCard.GetComponent<MemoryCard>().Unreveal();
                //tempCard.GetComponent<MemoryCard>().Reveal();
                //tempCard.GetComponent<SpriteRenderer>().sprite =
                images[numbers[i*gridCols+j]];
                float posX = (offsetX * j ) + originalX;
                float posY = (offsetY * i ) + originalY;
                tempCard.transform.position = new Vector3(posX,posY,0);
            }
        }
    }
    ……
    public bool CardReveal()
    {
        return secondRevealedCard == null;
    }
    //卡牌的翻转
    public void CardReveal(MemoryCard card)
    {
        if(firstRevealedCard == null)
```

```
        {//如果第 1 张卡牌对象为空，则把卡牌赋值给 firstRevealedCard
            firstRevealedCard = card;
            Debug.Log("FirstRevealedCard ID = " + firstRevealedCard.getID());
        }
        else
        {//如果第 1 张卡牌对象已经赋值，则把卡牌赋值给 secondRevealedCard
            secondRevealedCard = card;
            Debug.Log("SecondRevealedCard ID = " + secondRevealedCard.getID());
            StartCoroutine(CheckMatch());
        }
    }

    private IEnumerator CheckMatch()
    {
        if(firstRevealedCard.getID() == secondRevealedCard.getID())
        {//如果翻开的两张卡牌是配对的
            firstRevealedCard.Matched();                     //则将卡牌设置为配对成功状态
            secondRevealedCard.Matched();
        }
        else
        {
            yield return new WaitForSeconds(1.5f);           //等待 1.5 秒
            firstRevealedCard.Unreveal();                    //将卡牌设置成未被翻开状态
            secondRevealedCard.Unreveal();
        }
        firstRevealedCard = null;
        secondRevealedCard = null;
        //在两两选择后，将对象清空，以免影响下一次配对
    }
}
```

3.4.7 计算分数与步数

（1）创建文本框。在 Hierarchy 窗口中创建两个 TextMeshPro：一个重命名为 ScoreText，在文本框中输入"分数:"，并将其设置为白色；另一个重命名为 StepText，在文本框中输入"步数:"，并将其设置为白色。这两个文本框用于显示游戏的分数和步数。图 3-14 所示为分数与步数的显示效果。

图 3-14　分数与步数的显示效果

（2）管理分数与步数。在 SceneManager 脚本中定义 scoreText 对象和 stepText 对象，并

声明私有变量 score（分数）和 step（步数）。在单击两次后增加步数值，在配对成功后增加分数值。SceneManager 脚本中的代码如下：

```csharp
using UnityEngine.UI;                          //需要用到UnityEngine.UI 包
using TMPro;                                    //需要用到TMPro 包
public class SceneManager : MonoBehaviour
{
    ……
    public TextMeshProUGUI scoreText;           //分数文本
    public TextMeshProUGUI stepText;            //步数文本
    private int score;                          //设置记录分数的变量
    private int step;                           //设置记录步数的变量
    ……
    public void CardReveal(MemoryCard card)
    {
        if(firstRevealedCard == null)
        {
            firstRevealedCard = card;
            Debug.Log("FirstRevealedCard ID = " + firstRevealedCard.getID());
        }
        else{
            secondRevealedCard = card;
            Debug.Log("SecondRevealedCard ID = " + secondRevealedCard.getID());
            step++;
            stepText.text = "步数: " + step;     //在 stepText 文本上显示步数
            StartCoroutine(CheckMatch());
        }
    }
    ……
    private IEnumerator CheckMatch()
    {
        if(firstRevealedCard.getID() == secondRevealedCard.getID())
        {
            firstRevealedCard.Matched();
            secondRevealedCard.Matched();
            score++;
            scoreText.text = "分数: " + score;   //在 scoreText 文本上显示分数
        }
        else{
            yield return new WaitForSeconds(1.5f);
            firstRevealedCard.Unreveal();
            secondRevealedCard.Unreveal();
        }
        firstRevealedCard = null;
        secondRevealedCard = null;
    }
}
```

（3）在 Unity 界面中添加两个 TextMeshPro 对象，分别命名为 ScoreText 和 StepText。把以上两个 TextMeshPro 对象赋值给 SceneManager 脚本中的 Score Text 变量和 Step Text 变量，如图 3-15 所示。

（4）游戏成功界面。为了保证游戏的完整性，可以增加游戏结束判断。在 Canvas 中新建一个 Image，并命名为 VictoryImage。将 victory 图片赋值给 VictoryImage，如图 3-16 所示，同时将该 Image 隐藏。

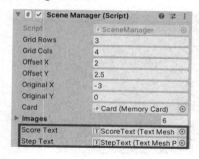

图 3-15 拖入对象

图 3-16 游戏成功界面

（5）游戏结束。victory 图片只在玩家完成游戏后才显示。SceneManager 脚本中的代码如下：

```
public class SceneManager : MonoBehaviour
{
    ……
    public Image victoryImage;           //victory 图片
    ……
    private IEnumerator CheckMatch()
    {
        if(firstRevealedCard.getID() == secondRevealedCard.getID())
        {
            firstRevealedCard.Matched();
            secondRevealedCard.Matched();
            score++;
            scoreText.text = "分数: " + score;
            Victory();    //游戏胜利
        }
        else{
            yield return new WaitForSeconds(1.5f);
            firstRevealedCard.Unreveal();
            secondRevealedCard.Unreveal();
        }
        firstRevealedCard = null;
        secondRevealedCard = null;
    }
    //游戏胜利
    void Victory()
    {
        if(score == (gridRows * gridCols) / 2)       //如果得分是卡牌总数的二分之一
        {
            victoryImage.gameObject.SetActive(true);  //显示游戏胜利界面
        }
    }
}
```

（6）拖入图片。在编写完代码以后不要忘记将 VictoryImage 拖入 SceneManager 脚本中，如图 3-17 所示。

（7）重新开始。先创建一个 Button 控件并将其放置在合适的位置，删除 text；再将图片

start-button 赋值给 Button 控件。为该 Button 控件新建并绑定一个新的脚本文件 Restart 后，双击打开该脚本文件，如图 3-18 所示。

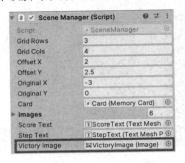

图 3-17 拖入 VictoryImage

图 3-18 重新开始

Restart 脚本中的代码如下：

```
public class Restart : MonoBehaviour
{
    public void RestartGame()    //重新开始游戏
    {
        //重载当前游戏场景
        UnityEngine.SceneManagement.SceneManager.LoadScene(0);
    }
}
```

3.5 本章小结

本章主要介绍了《记忆翻牌》游戏的设计与制作，其中重点在于卡牌结构的设计、洗牌算法的实现、卡牌的显示与配对等。读者需要灵活运用算法中涉及的数组等基本概念。

3.6 练习题

1. 本章介绍的简单洗牌算法（Shuffle）的空间复杂度与时间复杂度分别为多少？（假设有 n 张卡牌）
2. 在《记忆翻牌》游戏中，假设有 m 行 n 列的卡牌待翻转，卡牌牌面值在棋盘中服从均匀分布，且翻转卡牌的策略如下：从上一次成功翻出一对卡牌后，在本次成功翻出一对卡牌之前，保持第奇数次翻的卡牌不变，只用第偶数次翻的卡牌进行试探；沿用此策略翻转卡牌，直至最终全部匹配成功。在已翻转第 1 张卡牌后，翻转第 2 张与第 1 张匹配成功的概率为多少？
3. 请设计一种你觉得最优的翻转卡牌策略，使成功完成游戏的平均步数最少。

第 4 章

拼图

4.1 游戏简介

《拼图》是一款益智类经典游戏，其历史可以追溯至 1760 年左右。当时，英国的雕刻师和制图师 John Spilsbury 将地图镶嵌在硬木板上，并沿着国界线进行切割，由此打造出最原始的拼图。《拼图》游戏常具有教育意义，所使用的图片往往附有适合青少年阅读的短文，或者历史、地理知识。

4.2 游戏规则

将一张图片切割成无数小块，这些小块被称为"碎片"。玩家凭借智慧与耐心，将这些碎片逐一拼接起来，如还原出原始图片的样式，即可视为游戏成功。在这个过程中，玩家不仅要有足够的耐心，还需要具备一定的观察力和判断力，以便精准识别每个碎片的位置和用途。

4.3 游戏开发核心思路

4.3.1 碎片对象的生成与位置初始化

使用 Resources 类进行碎片资源的读入与对象生成，并保存在一个数组中。碎片资源如图 4-1 所示。

图 4-1　碎片资源

4.3.2 原图与碎片位置的对应关系

以 3×3 九宫格为例，可以将构成原图的 9 个碎片用不同的整数（如 1、2、3、4、5、6、7、8 和 9）来表示，并保存到一个 3×3 的二维数组中。

假设二维数组的第 1 个元素[0,0]保存第 1 个碎片，最后一个元素[2,2]保存最后一个碎片。二维数组中的每个碎片对象对应一维数组碎片的索引号可以表示为如下形式：

$$\text{图片一维数组索引号} = row \times colNumber + col$$

其中，row 表示二维数组中一个元素所在的行号，colNumber 表示二维数组每行的列数，col 表示二维数组中一个元素所在的列号。例如，[1,2]对应的碎片需要获取的是一维数组中索引号为 1×3+2 = 5 的图片。

假设碎片的宽度为 w，高度为 h，碎片的中心点位置为锚点，原图的左下角坐标为世界坐标的原点，即（0,0）的位置，假设当前碎片在二维数组中的索引号为[i][j]，那么计算当前碎片的位置的公式如下：

$$\begin{cases} x = \left(i + \dfrac{1}{2}\right) \times w \\ y = \left(j + \dfrac{1}{2}\right) \times h \end{cases}$$

例如，索引号为[2][2]的坐标可以根据上述公式求出为 $\left(\dfrac{5}{2}w, \dfrac{5}{2}h\right)$，如图 4-2 所示。

如果原图的中心为世界坐标系的原点，如图 4-3 所示，即（0,0）的位置，那么索引号为[i][j]的碎片的位置的计算公式如下：

$$\begin{cases} x = (j - 1) \times w \\ y = (i - 1) \times h \end{cases}$$

例如，索引号为[2][2]的碎片的位置为（w,h）。

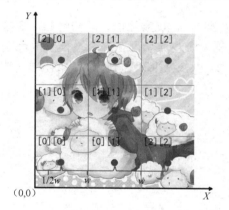

图 4-2　坐标原点在图片的左下角

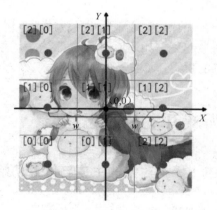

图 4-3　坐标原点在图片中心

4.3.3　拖动碎片

下面介绍通过鼠标拖动碎片的基本算法。

（1）判断单击的位置是否在碎片内。

① 把鼠标指针的屏幕坐标转换为世界坐标（Mx,My）。

② 获取碎片的坐标（Sx,Sy）。

③ 如果|Mx－Sx| < w/2, |My－Sy| < h/2，则表明鼠标指针的世界坐标位置在此碎片的范围内，表示碎片被选中（此处不考虑图层深度）。

（2）实现被拖动的碎片跟随鼠标指针移动，如图 4-4 所示。
① 获得鼠标指针的屏幕坐标位置，并转换为世界坐标的位置。

图 4-4 拖动碎片

② 碎片新的坐标 = 鼠标指针的世界坐标 + 碎片位置与鼠标指针位置的偏移量。
（3）当松开鼠标按键时，进行碎片位置是否匹配的判断。

4.3.4 碎片放置位置正确性的判断

通过计算碎片与正确位置的距离来判断碎片是否靠近了正确位置，$A(x_1,y_1)$ 和 $B(x_2,y_2)$ 分别表示碎片和对应的正确位置的坐标，如图 4-5 所示。

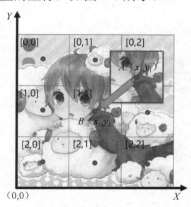

图 4-5 判断碎片是否在正确位置

在计算两点之间的距离时，有多种算法可以选择，典型算法有如下几种。
- 利用距离公式计算：$|AB| = \sqrt{(x_1 - x_2)^2 + (y_1 - y_2)^2}$。
- 利用距离公式的平方计算：$|AB|^2 = (x_1 - x_2)^2 + (y_1 - y_2)^2$。
- 使用绝对值和临界值进行判断：$|x_2 - x_1| < 临界值 < |y_1 - y_2|$。

可以使用更简单的方法来判断碎片是否接近正确位置，即获取碎片的坐标（Sx,Sy）与碎片对应的正确位置的坐标（Tx,Ty），若"|Tx − Sx|<临界值&&|Ty − Sy|<临界值"，则判定碎片靠近对应的正确位置，将碎片的坐标修改为（Tx,Ty）。若没有靠近，则将碎片返回初始位置。伪代码如下：

```
if ( |Tx - Sx| < 临界值 && |Ty - Sy| < 临界值){
    碎片坐标 = 正确坐标
}else{
    碎片回到原来的位置;
}
```

4.3.5 游戏流程图

游戏流程图如图 4-6 所示。

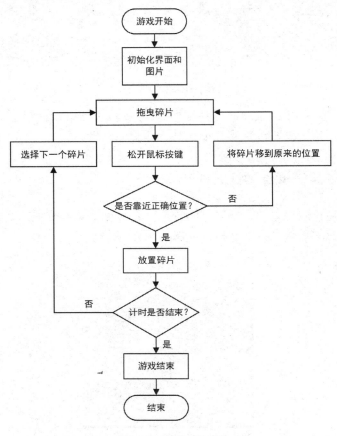

图 4-6 游戏流程图

4.4 游戏实现

4.4.1 新建工程与导入资源

（1）新建工程。新建名称为 Puzzle 的工程，并选择 2D 选项，如图 4-7 所示。

（2）导入资源包。选择 Assets→Import Package→Custom Package 命令，导入资源包 puzzle.unitypackage，如图 4-8 所示。

（3）设置摄像机的位置。在 Hierarchy 窗口中单击 Main Camera 物体，设置 Inspector 窗

口中的参数。调整摄像机的坐标，并调整背景的位置，使背景缩放并覆盖屏幕，如图 4-9 所示。

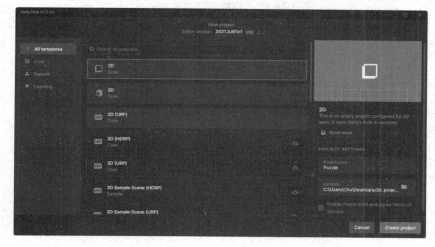

图 4-7　新建工程

图 4-8　导入资源包

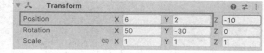

图 4-9　设置摄像机的位置

（4）设置原图。单击 Sprite Editor 按钮，将锚点拖到左下角。在修改完锚点的位置后，将原图左下角的坐标设置为（0,0）。调整图片的颜色，使其变暗一些。设置 Order in Layer 为 1，效果如图 4-10 所示。

（5）设置碎片。复制一个碎片并将其重命名为 patches，将 patches 的 Sprite Mode 设置为 Multiply，如图 4-11 所示，单击 Apply 按钮。

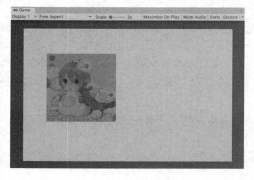

图 4-10　设置原图

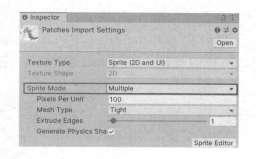

图 4-11　设置碎片

（6）切割图片。先单击 Sprite Editor 按钮，再单击 Slice 下拉按钮，将 Type 设置为 Grid By Cell Count，将 Column & Row 设置为 3 行 3 列，单击 Slice 按钮进行切割，如图 4-12 所示。

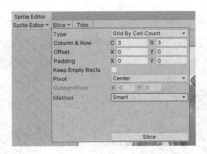

图 4-12 切割图片

4.4.2 批量读取碎片

（1）在 Scripts 文件夹中新建一个脚本并将其命名为 CreatePatch，此脚本用来生成碎片。可以通过代码来读取碎片的位置，并将碎片以数组的形式保存起来。CreatePatch 脚本中的代码如下：

```
public class CreatePatch : MonoBehaviour
{
    //读取碎片资源的路径
    public string patchFilePath = "Sprites/Patches";
    //存储碎片的数组
    private Sprite[] spritePatches;

    void Start()
    {
        LoadPatch(patchFilePath);
    }

    //根据路径批量读取碎片
    private void LoadPatch(string patchFilePath)
    {
        spritePatches = Resources.LoadAll<Sprite>(patchFilePath);
        foreach(Sprite sp in spritePatches)
        {
            Debug.Log(sp.ToString());
        }
    }
}
```

（2）新建一个空物体并将其命名为 GameManager。将 CreatePatch 脚本赋给 GameManager。运行游戏，可以看到，在 Console 窗口中碎片都已经被存储在数组中。图 4-13 所示为碎片成功读入的结果。

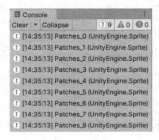

图 4-13 碎片成功读入的结果

4.4.3 生成碎片对象

返回 CreatePatch 脚本,引入一个二维数组,通过改变行和列的值设置该数组的大小。使用该数组来存储生成的碎片,同时将数组编号与生成的碎片编号一一对应。CreatePatch 脚本中的代码如下:

```csharp
public class CreatePatch : MonoBehaviour
{
    public string patchFilePath = "Sprites/Patches";
    private Sprite[] spritePatches;

    //存储碎片的二维数组
    private GameObject[,] patches;
    //碎片的行数
    private int rowNumber = 3;
    //碎片的列数
    private int colNumber = 3;

    void Start()
    {
        LoadPatch(patchFilePath);
        CreatePatchObj(spritePatches);
    }

    //创建拼图物体
    private void CreatePatchObj(Sprite[] sprite)
    {
        //创建一个二维数组存储碎片
        patches = new GameObject[rowNumber, colNumber];
        for (int row = 0; row < rowNumber; row++)
        {
            for (int col = 0; col < colNumber; col++)
            {
                //实例化一个空白碎片,并标上序号
                patches[row, col] = new GameObject("patch" + row + col);
                //设置碎片的 sprite 为对应碎片
                patches[row, col].AddComponent<SpriteRenderer>().sprite = sprite[
                                                    row* colNumber + col];
                //设置物体的图层
                patches[row, col].GetComponent<SpriteRenderer>().sortingOrder=2;
            }
        }
    }
}
```

4.4.4 初始化碎片位置

碎片的初始化放置只需要使用随机数函数 Random.Range()即可。
(1)返回 CreatePatch 脚本,在该脚本中添加代码。CreatePatch 脚本中的代码如下:

```csharp
public class CreatePatch : MonoBehaviour
{
```

```
//……

//生成碎片位置的X轴范围和Y轴范围
public Vector2 patchXRange, patchYRange;

//……
private void CreatePatchObj(Sprite[] sprite)
{
    patches = new GameObject[rowNumber, colNumber];
    for (int row = 0; row < rowNumber; row++)
    {
        for (int col = 0; col < colNumber; col++)
        {
            //……
            //使碎片随机在给定的范围内生成
            patches[row, col].GetComponent<Transform>().position = new
                Vector3(Random.Range(patchXRange.x, patchXRange.y),
                    Random.Range(patchYRange.x, patchYRange.y), 0);
        }
    }
}
```

（2）在场景中，为 CreatePatch 脚本输入需要的数据（随机生成范围），如图 4-14 所示。

图 4-14　输入需要的数据（随机生成范围）

4.4.5　计算每个碎片的目标位置

（1）创建一个脚本并将其命名为 Patch，该脚本用来保存每个碎片的基本信息。Patch 脚本中的代码如下：

```
public class Patch : MonoBehaviour
{
    //碎片的宽度和高度
    private float patchWidth, patchHeight;
    //碎片的起始位置
    private Vector3 initPosition;
    //碎片的正确位置
    private Vector3 targetPosition;

    //初始化碎片的成员变量
    public void InitPatch(float patchWidth, float patchHeight, Vector3 initPosition,
Vector3 targetPosition)
    {
        this.patchWidth = patchWidth;
        this.patchHeight = patchHeight;
        this.initPosition = initPosition;
        this.targetPosition = targetPosition;
```

```
    }
    //获得碎片的正确位置
    public Vector3 GetTargetPosition()
    {
        return targetPosition;
    }
}
```

（2）返回 CreatePatch 脚本，在生成随机位置后，为碎片添加 Patch 组件。CreatePatch 脚本中的代码如下：

```
public class CreatePatch : MonoBehaviour
{
    //……
    private void CreatePatchObj(Sprite[] sprite)
    {
        patches = new GameObject[rowNumber, colNumber];
        for (int row = 0; row < rowNumber; row++)
        {
            for (int col = 0; col < colNumber; col++)
            {
                //……
                //获得图片边界框的大小
                Vector2 patchSize = patches[row, col].GetComponent
                                    <SpriteRenderer>().sprite.bounds.size;
                //为生成的碎片添加 Patch 组件，并计算碎片的正确位置
                patches[row, col].AddComponent<Patch>().InitPatch(
                    patchSize.x, patchSize.y, patches[row, col].GetComponent
                    <Transform>().position,
                    new Vector3((col + 0.5f) * patchSize.x,
                    (rowNumber - row - 0.5f) * patchSize.y, 0)
                );
            }
        }
    }
}
```

4.4.6　拖动鼠标

（1）创建一个脚本并将其命名为 MouseDrag，将该脚本挂载到 GameManager 上。实现鼠标左键事件，监听鼠标按键的按下事件、按住事件和松开事件，并使用 Input 类的对应函数实现按键监听功能。MouseDrag 脚本中的代码如下：

```
public class MouseDrag : MonoBehaviour
{
    void Update()
    {
        if (Input.GetMouseButtonDown(0))               //鼠标左键按下时监听
        {
            Debug.Log(selectedPatch.ToString());
        }
        if (Input.GetMouseButton(0))                   //鼠标左键按住时监听
        {
```

```
            Debug.Log("GetMouseButton");
        }
        if (Input.GetMouseButtonUp(0))             //鼠标左键松开时监听
        {
            Debug.Log("GetMouseButtonUp");
        }
    }
}
```

（2）选取碎片。在单击碎片后，选取被单击的碎片。先将鼠标指针的位置由屏幕坐标转化为世界坐标，再判断其位置是否在碎片内。MouseDrag 脚本中的代码如下：

```
public class MouseDrag : MonoBehaviour
{
    //CreatePatch 对象的引用
    public CreatePatch createPatchObj;
    //摄像机的引用
    public Camera camera;

    //世界坐标系下鼠标指针的位置
    private Vector3 mousePosInWorld;
    //当前接受判断的碎片
    private Patch selectingPatch;
    //当前被选择的碎片
    private Transform selectedPatch;

    void Update()
    {
        if (Input.GetMouseButtonDown(0))
        {
            //Debug.Log(selectedPatch.ToString());
            selectedPatch = SelectingPatch();      //返回选中的碎片
            if (selectedPatch != null)             //如果选中的碎片不为空
            {
                Debug.Log(selectedPatch.ToString());
            }
        }
        if (Input.GetMouseButton(0))
        {
            //Debug.Log("GetMouseButton");
        }
        if (Input.GetMouseButtonUp(0))
        {
            //Debug.Log("GetMouseButtonUp");
        }
    }

    //返回当前单击的碎片,若未选上碎片则返回 null
    private Transform SelectingPatch()
    {
        //获得鼠标指针的屏幕坐标,并将其转化为世界坐标
        mousePosInWorld = camera.ScreenToWorldPoint(Input.mousePosition);
        //循环所有碎片
        for (int i = 0; i < createPatchObj.GetRow(); i++)
```

```csharp
            {
                for (int j = 0; j < createPatchObj.GetCol(); j++)
                {
                    //将碎片从保存碎片的数组中取出
                    selectingPatch = createPatchObj.GetPatchObj()[i, j]
                            .GetComponent<Patch>();
                    if (selectingPatch.GetMatched() == false)
                    {
                        //判断鼠标指针的位置是否在该碎片内
                        if (Mathf.Abs(
                            mousePosInWorld.x - selectingPatch.transform.position.x)<
                    selectingPatch.GetWidth() / 2 &&
                            Mathf.Abs(
                            mousePosInWorld.y - selectingPatch.transform.position.y) <
                            selectingPatch.GetHeight() / 2)
                        {
                            return selectingPatch.gameObject.transform;
                        }
                    }
                }
            }
        return null;
    }
}
```

CreatePatch 脚本中的代码如下:

```csharp
public class CreatePatch : MonoBehaviour
{
    ……
    //声明一个函数来返回保存碎片的数组
    public GameObject[,] GetPatchObj()
    {
        return patches;
    }
    //返回二维数组的行、列值
    public int GetRow(){return rowNumber;}
    public int GetCol(){return colNumber;}
}
```

Patch 脚本中的代码如下:

```csharp
public class Patch : MonoBehaviour
{
    ……
    //返回碎片是否已经处于正确位置
    private bool isMatched = false;
    ……
    //返回碎片的宽度和高度
    public float GetWidth(){return patchWidth;}
    public float GetHeight(){return patchHeight;}
    //修改碎片是否已经处于正确位置
    public void SetMatched(){isMatched = true;}
    public bool GetMatched(){return isMatched;}
}
```

（3）移动碎片。在按住鼠标左键时，使被选择的碎片跟随鼠标指针移动。当 Input.GetMouseButton(0) 为 true 时，使碎片的位置加上鼠标指针每帧的位移。MouseDrag 脚本中的代码如下：

```csharp
public class MouseDrag : MonoBehaviour
{
    //……
    //上一帧鼠标指针的位置
    private Vector3 lastMousePosition = Vector3.zero;

    void Update()
    {
        if (Input.GetMouseButtonDown(0))
        {
            //……
        }
        if (Input.GetMouseButton(0))
        {
            //Debug.Log("GetMouseButton");
            DragPatch();   //拖动碎片
        }
        if (Input.GetMouseButtonUp(0))
        {
            //Debug.Log("GetMouseButtonUp");
            if (selectedPatch != null)
            {
                //设置碎片的图层
                selectedPatch.GetComponent<SpriteRenderer>().sortingOrder = 2;
            }
            //拖动结束，将鼠标指针的位置清零
            lastMousePosition = Vector3.zero;
        }
    }

    //……

    //使被选择碎片跟随鼠标指针移动
    private void DragPatch()
    {
        if (selectedPatch == null) return;
        if (lastMousePosition != Vector3.zero)
        {
            //更改被拖曳碎片的层级顺序，使其始终渲染于其他碎片之上
            selectedPatch.GetComponent<SpriteRenderer>().sortingOrder = 5;
            //计算鼠标指针每帧移动的距离
            Vector3 offset = camera.ScreenToWorldPoint(Input.mousePosition) -
                        lastMousePosition;
            //将碎片随着鼠标指针移动相应的距离
            selectedPatch.position += offset;
        }
        //更新上一帧鼠标指针的位置
        lastMousePosition = camera.ScreenToWorldPoint(Input.mousePosition);
    }
}
```

（4）碎片位置与目标位置的匹配。判断碎片是否已到达正确位置，如果没有到达正确位置，则让碎片回到原位。需要判断被选择碎片与其正确位置各自的 x 坐标和 y 坐标的差值是否小于某个阈值。MouseDrag 脚本中的代码如下：

```
public class MouseDrag : MonoBehaviour
{
    //……
    //配对碎片到达正确位置时的阈值
    public float threshold = 0.2f;

    //……

    //判断碎片是否已到达正确位置，如果没有到达正确位置，则让碎片回到原位
    private void MatchPatch()
    {
        if (selectedPatch == null) return;
        //获取选择的碎片
        Patch currentPatch = selectedPatch.GetComponent<Patch>();
        //如果碎片的坐标在阈值内
        if (Mathf.Abs(selectedPatch.position.x - currentPatch.GetTargetPosition().x) <
                threshold &&
            Mathf.Abs(selectedPatch.position.y - currentPatch.GetTargetPosition().y) <
                threshold)
        {
            //使碎片移到正确位置
            selectedPatch.position = currentPatch.GetTargetPosition();
            currentPatch.SetMatched();
        }
        else
        {
            //使碎片返回原位
            selectedPatch.position = currentPatch.GetInitPosition();
        }
    }
}
```

Patch 脚本中的代码如下：

```
public class Patch : MonoBehaviour
{
    //……
    //返回碎片生成时的坐标
    public Vector3 GetInitPosition() { return initPosition; }
}
```

（5）为脚本赋值，如图 4-15 所示。

图 4-15　为脚本赋值

4.4.7 判断游戏是否胜利

（1）创建一个脚本并将其命名为 GameOver。可以用分数来计算碎片配对成功的次数，如果分数等于碎片的个数，则所有碎片已经完成配对，此时游戏胜利。GameOver 脚本中的代码如下：

```
public class GameOver : MonoBehaviour
{
    //CreatePatch 对象的引用
    public CreatePatch createPatchObj;
    //已经到达正确位置的碎片数量
    private int matchedPatchCount = 0;
    //增加到达正确位置的碎片计数，当计数等于总碎片数时，游戏胜利
    public void Judge()
    {
        matchedPatchCount++;
        if (matchedPatchCount == createPatchObj.GetPatchCount())
        {
            Debug.Log("You Win!!!");
        }
    }
}
```

CreatePatch 脚本中的代码如下：

```
public class CreatePatch : MonoBehaviour
{
    //……
    //返回总碎片数
    public int GetPatchCount() { return rowNumber * colNumber; }
}
```

MouseDrag 脚本中的代码如下：

```
public class MouseDrag : MonoBehaviour
{
    //……
    //GameOver 对象的引用
    public GameOver gameOver;

    private void MatchPatch()
    {
        if (selectedPatch == null) return;
        Patch currentPatch = selectedPatch.GetComponent<Patch>();
        if (Mathf.Abs(selectedPatch.position.x - currentPatch.GetTargetPosition().x) <
            threshold &&
            Mathf.Abs(selectedPatch.position.y - currentPatch.GetTargetPosition().y) <
            threshold)
        {
            selectedPatch.position = currentPatch.GetTargetPosition();
            currentPatch.SetMatched();
            gameOver.Judge();    //执行胜利判断
        }
        else
        {
            selectedPatch.position = currentPatch.GetInitPosition();
```

            }
        }
    }

（2）将 GameOver 脚本挂载到 GameManager 上，把 CreatePatch 脚本赋值给 GameOver，同时把 GameOver 脚本赋值给 MouseDrag，如图 4-16 所示。

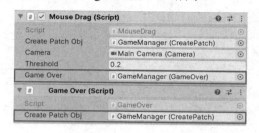

图 4-16　为 GameOver 和 MouseDrag 赋值

4.5　本章小结

本章主要介绍了《拼图》游戏的设计与制作，其中重点在于对游戏物体的各种坐标系的认识及应用，包括局部坐标系、世界坐标系。另外，读者应对各种距离有简单的认识，如常见的欧氏距离等。

4.6　练习题

1. 假设《拼图》游戏碎片的宽度和高度分别为 w 和 h，且其锚点定位在碎片中心。锚点在世界坐标系下的位置为 (u,v)，若要确保鼠标指针在碎片内部（不包括边缘），则鼠标指针在世界坐标系下的位置 (x,y) 应满足何种条件？

2. 假设《拼图》游戏碎片的宽度和高度分别为 w 和 h，且其锚点定位在原图左下角，原图左下角在世界坐标系的原点，则第 r 行第 c 列（r 属于 $[0,h]$ 区间，c 属于 $[0,w]$ 区间）碎片的目标位置的坐标为多少？

3.（选做）目前的《拼图》游戏有一个小问题，当碎片随机初始化时，前后遮挡关系随机，当单击重合区域时，选中的碎片不一定是前面的被遮挡还是后面的被遮挡的。请设计一种解决方案，使每次单击时一定会选中可见区域对应的碎片（而非被遮挡的碎片）。

第 5 章

推箱子

5.1 游戏简介

《推箱子》游戏是在一个正方形的棋盘上进行的，每个方块表示一块地板或一面墙。地板可以通过，墙不可以通过。地板上放置了箱子，一些地板被标记为存储位置。玩家被限制在棋盘上，可以水平或垂直地移动到空的方块上（永远不会穿过墙或箱子）。箱子不得被推入其他箱子或墙中，也不能被拉出。箱子的数量等于存储位置的数量。当所有的箱子都放在存储位置时，游戏胜利。《推箱子》游戏的画面如图 5-1 所示。

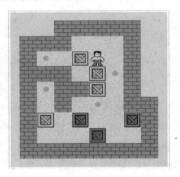

图 5-1 　《推箱子》游戏的画面

5.2 游戏规则

《推箱子》游戏就像是在一个巨大的魔幻棋盘上进行的一场冒险。在棋盘上，每个方块就像是一个小房间或迷宫，有的可以走进去，有的则设置了阻挡前进的墙。这些墙就像是一个个的关卡，玩家不能穿越它们，只能绕道而行。在地板上还摆放了一些箱子，它们散落在一个个小房间中，好像是一个个陷阱。有些房间被特别标记。玩家必须非常仔细地规划每一步，把所有的箱子安全地送到存储位置才能取得游戏的胜利。《推箱子》游戏的规则如图 5-2 所示。

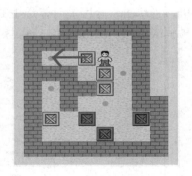

图 5-2 　《推箱子》游戏的规则

5.3 游戏开发核心思路

5.3.1 地图的生成

可以使用二维数组来存储地图信息，每个元素用不同的数字标记不同的对象。0 表示空白，1 表示墙，2 表示角色，3 表示箱子，9 表示终点（9 表示终点是为了方便以后添加其他颜色的箱子），如图 5-3 所示。

地图的生成是通过遍历地图数组和读取每个元素的数值来完成的。根据数值标记可以生成不同的地图对象。

假设坐标原点在地图的左上角，向右为 x 轴正方向，向下为 y 轴负方向。二维数组 map[m][n]第 i 行第 j 列的元素为 map[i][j]，每个方块的尺寸为 1×1，则对应 map[i][j]在场景中的坐标位置为 position = (j,-i)。例如，map[1][4]对应的方块的坐标位置为（4,-1）。同时，该元素的值为 1，表示的是墙的方块，如图 5-4 所示。

图 5-3 用于存储地图信息的二维数组

图 5-4 数组与地图相对应

为了使程序实现更简便，可以把地图的二维数组的数据映射到一维数组中，以便遍历。假设二维数组为 map[m][n]，一维数组为 b[m×n]，则它们的对应关系为 b[i×n+j] = map[i][j]。

假设方块的长和宽都为 1，地图二维数组的大小为 9×9，则二维数组中的 map[2][1]对应一维数组中的 b[2×9+1] = b[19]。也就是说，对于数组元素 b[19]，对应的实际的坐标为（1,-2）。

虽然一维数组对于我们来说不是很直观，但是把二维数组转换成一维数组有利于程序实现。因此，在设计地图时用二维数组来思考，在进行程序操作（如角色移动）时用一维数组来表示。

5.3.2 角色位置及移动方向与数组的对应关系

角色在数组中的表示。假设地图左上角为坐标原点，即 map[0][0]，对应一维数组中的 b[0]。若角色位于 map[i][j]，则对应一维数组中的 b[i×9+j]。此处以 map[4][4]（即 b[4×9+4]= b[40]）为例进行介绍，如图 5-5 所示。

- 角色上方一个方格的二维表示形式为 map[3][4]，对应的一维表示形式为 b[3×9+4] = b[4×9+4-9] =b[31]。
- 角色下方一个方格的二维表示形式为 map[5][4]，对应的一维表示形式为 b[5×9+4] = b[4×9+4+9] =b[49]。
- 角色左侧一个方格的二维表示形式为 map[4][3]，对应的一维表示形式为 b[4×9+3] = b[4×9+4-1] =b[39]。

- 角色右侧一个方格的二维表示形式为 map[4][5]，对应的一维表示形式为 $b[4×9+5] = b[4×9+4+1] = b[41]$。

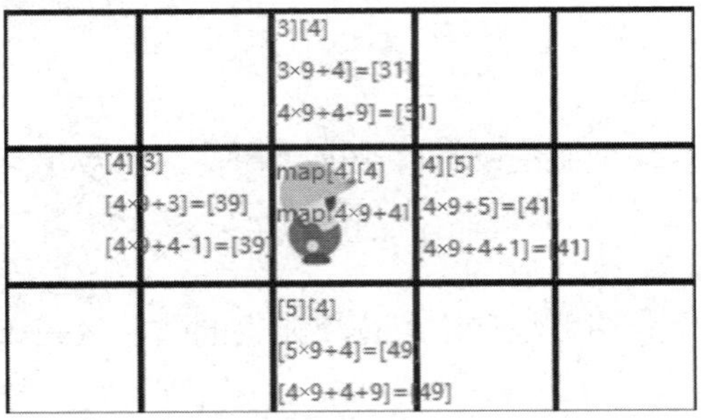

图 5-5　角色在数组中的表示

由此可以推断出以下几点。
- 角色往上移动一格，在一维数组中索引-9。
- 角色往下移动一格，在一维数组中索引+9。
- 角色往左移动一格，在一维数组中索引-1。
- 角色往右移动一格，在一维数组中索引+1。

因此，可以使用一个枚举类型来表示 4 个移动方向：public enum Direction { Up = -9,Down = 9,Left = -1,Right = 1}。

5.3.3　分析角色可移动和不可移动情况

角色可移动和不可移动情况如图 5-6 所示。可以看出，在《推箱子》游戏中，角色可移动和不可移动的状态比较多。

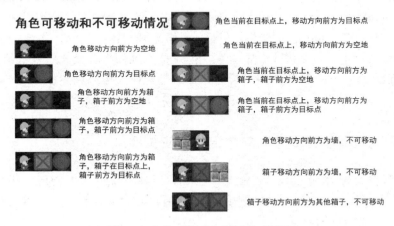

图 5-6　角色可移动和不可移动情况

5.3.4　角色移动的代码逻辑

角色移动如果采用 if-else 语句来判断，则代码逻辑混乱，不易修改与调试，且可读性差。因此，可以采用 switch 语句来判断，这样代码比较简洁，逻辑清晰，比较容易修改与调试，且可读性比较好。

1. 角色的移动与地图的刷新

- 向右移动：获取到按键"→"，此时 dir=1。将角色目前位置的数组值变为 0，下一个目标移动位置的数组值变为 2，刷新地图数组（游戏状态快照）。
- 向上移动：获取到按键"↑"，此时 dir=-9。将角色目前位置的数组值变为 0，下一个目标移动位置的数组值变为 2，刷新地图数组（游戏状态快照）。
- 向下移动：获取到按键"↓"，此时 dir=9。将角色目前位置的数组值变为 0，下一个目标移动位置的数组值变为 2，刷新地图数组（游戏状态快照）。
- 向左移动：获取到按键"←"，此时 dir=-1。将角色目前位置的数组值变为 0，下一个目标移动位置的数组值变为 2，刷新地图数组（游戏状态快照）。
- 当遇到障碍物时：跳出遍历，返回 false。

2. 箱子的移动与地图的刷新

箱子的移动与角色的移动算法类似，只是除刷新角色的位置外，还需要刷新箱子的位置。移动箱子的功能如下：设定 dir 控制角色在数组中移动的距离。

- 向右移动：获取到按键"→"，此时 dir=1。将角色目前位置的数组值变为 0，下一个目标移动位置的数组值变为 2，下下个目标移动位置的数组值变为 3，刷新地图数组（游戏状态快照）。
- 向上移动：获取到按键"↑"，此时 dir=-9。将角色目前位置的数组值变为 0，下一个目标移动位置的数组值变为 2，下下个目标移动位置的数组值变为 3，刷新地图数组（游戏状态快照）。
- 向下移动：获取到按键"↓"，此时 dir=9。将角色目前位置的数组值变为 0，下一个目标移动位置的数组值变为 2，下下个目标移动位置的数组值变为 3，刷新地图数组（游戏状态快照）。
- 向左移动：获取到按键"←"，此时 dir=-1。将角色目前位置的数组值变为 0，下一个目标移动位置的数组值变为 2，下下个目标移动位置的数组值变为 3，刷新地图数组（游戏状态快照）。
- 当遇到障碍物时：跳出遍历，返回 false。

角色移动的伪代码框架如下：

```
switch（角色前方位置状态）
{
    case 前方为空：
    break;
    case 前方为目标点：
    break;
    case 前方为箱子：
        switch（箱子前方状态）
        {
            case 前方为空：
            break;
            case 前方为目标点：
            break;
        }
    break;
    case 前方为箱子在目标点上：
        switch（箱子前方状态）：
```

```
            {
                case 前方为空:
                    break;
                case 前方为目标点:
                    break;
            }
        break;
    default:
        不可移动;
        break;
}
```

5.3.5 游戏胜利的判定

创建一个计数变量,并设置为当前关卡中目标点的数量。当箱子被移动到目标点时,该变量的值-1。当箱子被移出目标点时,该变量的值+1。当该变量的值为 0 时,游戏结束。

5.3.6 游戏流程图

游戏流程图如图 5-7 所示。

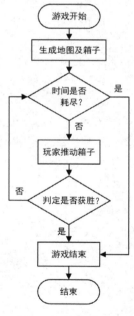

图 5-7　游戏流程图

5.4 游戏实现

5.4.1 资源的导入与切割

(1) 新建工程。单击新项目按钮,选择 2D 选项,新建一个名称为 Sokoban 的工程,如

图 5-8 所示。

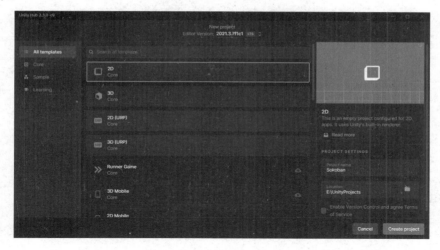

图 5-8 新建工程

（2）导入资源。导入资源包 Sokoban，该资源包中有完整的游戏实例和完成游戏所需的所有素材。

（3）设置 Sprite Mode。选择 Tiles 图片，将 Sprite Mode 设置为 Multiple，如图 5-9 所示，单击 Apply 按钮。

（4）切割素材。先单击 Sprite Editor 按钮，再单击 Slice 下拉按钮，将 Type 设置为 Grid By Cell Size。因为 Tiles 图片是 256 像素×256 像素的，每个素材是 64 像素×64 像素的，所以把它切割为 64 像素×64 像素，单击 Slice 按钮，如图 5-10 所示。

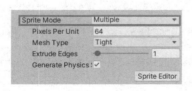

图 5-9 设置 Sprite Mode

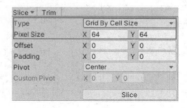

图 5-10 切割素材

（5）为素材命名。单击切割完的单个素材，修改右下角的素材名，如图 5-11 所示。

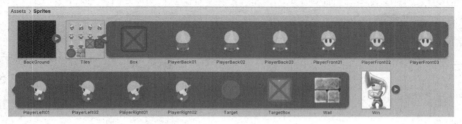

图 5-11 为素材命名

（6）设置游戏场景。将游戏背景拖入场景中，并将其坐标设置为（-4,4），同时将摄像机的坐标也设置为（-4,4）。将游戏背景的 Order in Layer 设置为-1，使游戏背景位于最下层，如图 5-12 所示。

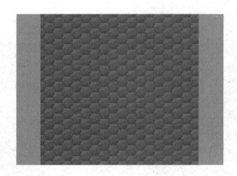

图 5-12　设置游戏场景

5.4.2　角色动画的制作

（1）修改角色图片的尺寸。选择 Tiles 图片，将 Pixels Per Unit 设置为 64，如图 5-13 所示，Pixels Per Unit 表示每个单位总共有多少像素。

（2）创建角色动画。将 PlayerFront01 图片拖入场景中，并将其名称修改为 Player。选择 Window→Animation 命令。在打开的 Animation 窗口中，单击 Create 按钮，创建一个新的动画序列。新建 Animation 文件夹，并在该文件夹中保存动画，将该动画序列命名为 FrontAnim。单击 Animation 窗口中右上角的图标，在弹出的菜单中选择 Show Sample Rate 命令，如图 5-14 所示，将 Samples 的值修改为 12（Samples 代表每秒钟播放的帧数）。

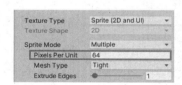

图 5-13　修改角色图片的尺寸　　　　　　图 5-14　创建角色动画

（3）添加动画。先将 front 图片一帧一帧地拖到动画序列中，如图 5-15 所示，再制作其他动画。单击 FrontAnim 下拉按钮，在弹出的下拉菜单中选择 Create New Clip 命令，按照上述方法制作 BackAnim 动画、LeftAnim 动画和 RightAnim 动画。

（4）设置动画状态机。单击 Player，选择 Window→Animator 命令，打开 Animation 窗口。选中 Any State，单击鼠标右键，在弹出的快捷菜单中选择 Make Transition 命令，出现一个箭头，将 4 个状态都和 Any State 连接起来，如图 5-16 所示。Any State 表示任何状态，连接后表示在任何状态下都可以转换为这 4 个动画。

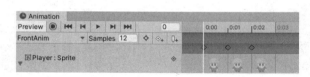
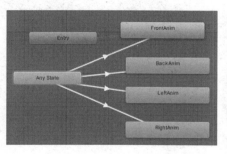

图 5-15　添加动画　　　　　　　　　　　图 5-16　设置动画状态机

（5）保存预制体。将 Player 保存为预制体，并删除场景中的 Player。

5.4.3 地图生成的程序实现与代码重构

（1）创建脚本并将其命名为 MapBuilder。需要根据设定的数据自动生成地图，所以可以使用一维数组存储地图信息，即将二维数组映射为一维数组，使用不同的数字表示不同的游戏物体。MapBuilder 脚本中的代码如下：

```csharp
public class MapBuilder : MonoBehaviour
{
    //地图的行数和列数
    public int rowNumber = 9, colNumber = 9;
    //地图数组（地图快照）
    private int[] snapshotMap;
    //墙体预制体
    public GameObject wall;
    //玩家预制体
    public GameObject player;
    //箱子预制体
    public GameObject box;
    //目标点预制体
    public GameObject targetPoint;
    //地板类型枚举
    public enum TileType{Null = 0, Wall = 1, Player = 2, Box = 3, TargetPoint = 9}

    void Start()
    {
        InitMap();
        BuildMap();
    }

    //初始化地图数组
    private void InitMap()
    {
        boxObjects = new GameObject[rowNumber*colNumber];
        snapshotMap = new int[]
        {
            1, 1, 1, 1, 1, 0, 0, 0, 0,
            1, 2, 0, 0, 1, 0, 0, 0, 0,
            1, 0, 3, 0, 1, 0, 1, 1, 1,
            1, 0, 3, 0, 1, 0, 1, 9, 1,
            1, 1, 1, 3, 1, 1, 1, 9, 1,
            0, 1, 1, 0, 0, 0, 0, 9, 1,
            0, 1, 0, 0, 0, 1, 0, 0, 1,
            0, 1, 0, 0, 0, 1, 1, 1, 1,
            0, 1, 1, 1, 1, 1, 0, 0, 0
        };
    }

    //根据地图数组生成地图
    private void BuildMap()
    {
```

```
        //循环二维数组
        for(int row = 0; row < rowNumber; row++)
        {
            for(int col = 0; col < colNumber;col++)
            {
                switch(snapshotMap[row * rowNumber + col])
                {
                    case (int)TileType.Null:
                        break;
                    case (int)TileType.Wall:
                        BuildWall(row,col);          //生成墙
                        break;
                    case (int)TileType.Player:
                        BuildPlayer(row,col);        //生成玩家
                        break;
                    case (int)TileType.TargetPoint:
                        BuildTargetPoint(row,col);   //生成目标点
                        break;
                }
            }
        }
    }
    //生成墙物体
    private void BuildWall(int row,int col)
    {
        GameObject wallObj = Instantiate(wall) as GameObject;
        wallObj.transform.position = new Vector3(col, -row, 0);
    }
    //生成玩家物体
    private void BuildPlayer(int row,int col)
    {
        GameObject playerObj = Instantiate(player) as GameObject;
        playerObj.transform.position = new Vector3(col, -row, 0);
    }
    //生成目标点物体
    private void BuildTargetPoint (int row,int col)
    {
        GameObject targetPointObj = Instantiate(targetPoint) as GameObject;
        targetPointObj.transform.position = new Vector3(col, -row, 0);
    }
}
```

（2）在场景中，创建一个新的空游戏对象并将其命名为 GameController，将 MapBuilder 脚本挂载到该对象上。将 Wall 和 Target Point 保存为预制体，并与 Player 预制体一起赋值给 MapBuilder 脚本，如图 5-17 所示。

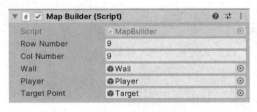

图 5-17 挂载 MapBuilder 脚本

（3）优化代码。通过观察代码可以发现，BuildWall()函数、BuildPlayer()函数和 BuildTargetPoint()函数除了参数不同，代码逻辑都是相同的。所以，需要重构函数，对代码进行修改，提高代码的可读性。MapBuilder 脚本中的代码如下：

```csharp
public class MapBuilder : MonoBehaviour
{
    ……
    private void BuildMap()
    {
        for(int row = 0; row < rowNumber; row++)
        {
            for(int col = 0; col < colNumber;col++)
            {
                switch(snapshotMap[row * rowNumber + col])
                {
                    case (int)TileType.Null:
                        break;
                    case (int)TileType.Wall:
                        BuildTile(row, col, wall);          //生成墙
                        break;
                    case (int)TileType.Player:
                        BuildTile(row,col,player);          //生成玩家
                        break;
                    case (int)TileType.TargetPoint:
                        BuildTile(row,col,targetPoint);     //生成目标点
                        break;
                }
            }
        }
    }

    //在地图指定位置生成游戏物体
    private GameObject BuildTile(int row, int col , GameObject tile)
    {
        GameObject tileObj = Instantiate(tile) as GameObject;
        tileObj.transform.position = new Vector3(col,-row,0);
        tileObj.name = tile.name;
        return tileObj;
    }
}
```

5.4.4　角色移动方向位置上的 Tile 类型检测与实现

（1）创建一个新的脚本并将其命名为 GameController。在玩家按下方向键后，需要检查对应方向上将要移动的 Tile 类型。可以将方向映射为一维数组的索引来进行加减，从而判断该方向上的 Tile 类型。GameController 脚本中的代码如下：

```csharp
public class GameController : MonoBehaviour
{
    //移动方向的枚举
    public enum Direction{Null = 0, Up = -9, Down = 9,Left = -1,Right = 1}
    //移动方向
```

```csharp
    private Direction dir;
    //MapBuilder 对象的引用
    public MapBuilder map;
    //玩家的 Transform 组件
    public Transform playerTransform;

    void Update()
    {
        DetectPlayer();
        DetectInput();
        MovePlayer();
    }

    //获得玩家角色的组件
    private void DetectPlayer()
    {
        if(playerTransform == null)
        {
            playerTransform = GameObject.Find("Player").GetComponent<Transform>();
        }
    }

    //方向键输入检测
    private void DetectInput()
    {
        //向上
        if(Input.GetKeyDown(KeyCode.UpArrow)) { dir = Direction.Up;}
        //向下
        else if(Input.GetKeyDown(KeyCode.DownArrow)) { dir = Direction.Down;}
        //向左
        else if(Input.GetKeyDown(KeyCode.LeftArrow)) { dir = Direction.Left;}
        //向右
        else if(Input.GetKeyDown(KeyCode.RightArrow)) { dir = Direction.Right;}
        else{dir = Direction.Null;}
    }

    //移动玩家角色
    private void MovePlayer()
    {
        switch(dir)
        {
            //向上
            case Direction.Up:
                Debug.Log("Up = " + map.GetTileType(playerTransform.position,dir));
                break;
            //向下
            case Direction.Down:
                Debug.Log("Down = " + map.GetTileType(playerTransform.position,dir));
                break;
            //向左
            case Direction.Left:
                Debug.Log("Left = " + map.GetTileType(playerTransform.position,dir));
```

```
            break;
        //向右
        case Direction.Right:
            Debug.Log("Right = " + map.GetTileType(playerTransform.position,dir));
            break;
        case Direction.Null:
            Debug.Log("Player = "+map.GetTileType(playerTransform.position,dir));
            break;
    }
}
```

修改 MapBuilder 脚本。修改后 MapBuilder 脚本中的代码如下：

```
public class MapBuilder : MonoBehaviour
{
    ……
    private void BuildMap()
    {
        for(int row = 0; row < rowNumber; row++)
        {
            for(int col = 0; col < colNumber;col++)
            {
                switch(snapshotMap[row * rowNumber + col])
                {
                    ……
                    //生成箱子
                    case (int)TileType.Box:
                        BuildTile(row,col,box);
                        break;
                }
            }
        }
    }
}
```

（2）将 GameController 脚本挂载到 GameController 对象上，并为脚本赋值，如图 5-18 所示。

图 5-18　挂载 GameController 脚本

5.4.5　角色可移动情况的代码框架实现

在 GameController 脚本和 MapBuilder 脚本中，需要检测角色可移动情况，可以使用 switch 语句进行判断。MapBuilder 脚本中的代码如下：

```
public class MapBuilder : MonoBehaviour
{
    ……
    //添加两个地板类型枚举
    public enum TileType{Null = 0, Wall = 1, Player = 2, Box = 3,
```

```
                            TargetPoint = 9, TargetWithBox = 10, TargetWithPlayer = 11}
    ……
}
```

GameController 脚本中的代码如下：

```
public class GameController : MonoBehaviour
{
    ……
    //角色的位置
    private Vector3 playerPos = Vector3.zero;
    ……
    private void DetectPlayer()
    {
        if(playerTransform == null)
        {
            playerTransform = GameObject.Find("Player").GetComponent<Transform>();
        }
        playerPos = playerTransform.position;  //获得角色当前的位置
    }

    private void Judge()
    {
        switch(map.GetTileType(playerPos,dir))
        {
            //角色移动方向的下一个位置为空
            case MapBuilder.TileType.Null:
                break;
            //角色移动方向的下一个位置为目标点
            case MapBuilder.TileType.TargetPoint:
                break;
            //角色移动方向的下一个位置为箱子
            case MapBuilder.TileType.Box:
                switch(map.GetTileType(playerPos,(Direction)((int)dir*2)))
                {
                    //箱子的下一个位置为空
                    case MapBuilder.TileType.Null:
                        break;
                    //箱子的下一个位置为目标点
                    case MapBuilder.TileType.TargetPoint:
                        break;
                }
                break;
            //角色移动方向的下一个位置为箱子，箱子在目标点上
            case MapBuilder.TileType.TargetWithBox:
                switch(map.GetTileType(playerPos,(Direction)((int)dir*2)))
                {
                    //箱子的下一个位置为空
                    case MapBuilder.TileType.Null:
                        break;
                    //箱子的下一个位置为目标点
                    case MapBuilder.TileType.TargetPoint:
                        break;
                }
                break;
```

```
            //其他情况
            default:
                break;
        }
    }
}
```

5.4.6 角色在场景中的移动功能

在获取到移动角色的命令后,变更角色当前位置的数组值,并改变下一个目标移动位置的数组值(变为代表角色的数值)。在变更完后,刷新地图数组,但是当遇到障碍物时,应该跳出遍历,停止角色移动,同时需要考虑到箱子的移动(可以依据角色移动进行推导)。当按方向键移动角色时,应根据按键方向将角色的位置加上对应的方向向量。GameController 脚本中的代码如下:

```
public class GameController : MonoBehaviour
{
    ......
    //角色移动的目标位置
    private Vector3 nextPlayerPos = Vector3.zero;
    //角色是否可以移动
    private bool canMovePlayer = false;
    ......
    void Update(
    {
        DetectPlayer();
        DetectInput();
        Judge();
        MovePlayer();
    }
    ......
    private void MovePlayer()
    {
        switch(dir)
        {
        case Direction.Up:
            nextPlayerPos = new Vector3(0,1);           //改变角色的位置,向上移动
            break;
        case Direction.Down:
            nextPlayerPos = new Vector3(0,-1);          //改变角色的位置,向下移动
            break;
        case Direction.Left:
            nextPlayerPos = new Vector3(-1,0);          //改变角色的位置,向左移动
            break;
        case Direction.Right:
            nextPlayerPos = new Vector3(1,0);           //改变角色的位置,向右移动
            break;
        case Direction.Null:
            nextPlayerPos = new Vector3(0,0);           //不改变角色的位置
            break;
        }
        if(canMovePlayer)
```

```csharp
            playerTransform.position += nextPlayerPos;        //改变角色的位置
        }
    }

    private void Judge()
    {
        switch(map.GetTileType(playerPos,dir))
        {
        case MapBuilder.TileType.Null:
            canMovePlayer = true;     //角色可以移动
            break;
        case MapBuilder.TileType.TargetPoint:
            canMovePlayer = true;     //角色可以移动
            break;
        case MapBuilder.TileType.Box:
            switch(map.GetTileType(playerPos,(Direction)((int)dir*2)))
            {
            ……
            }
            break;
        case MapBuilder.TileType.TargetWithBox:
            switch(map.GetTileType(playerPos,(Direction)((int)dir*2)))
            {
            ……
            }
            break;
        default:
            Debug.Log("Can`t move");
            canMovePlayer = false;    //角色不能移动
            break;
        }
    }
}
```

5.4.7 打印地图快照信息

（1）为了方便查看地图信息，可以使用一个 Text 来打印地图信息。先在场景中新建 Canvas 和 Text，将 Text 的显示字符修改为 0，并将字体颜色修改为黑色或红色。接着将 Text 保存为预制体，并将场景中的 Text 删除。

（2）可以用代码控制在 Canvas 下批量生成 Text。GameController 脚本中的代码如下：

```csharp
using UnityEngine.UI;

public class GameController : MonoBehaviour
{
    ……
    //显示 Tile 类型的 Text
    public Text tileTypeText;
    //UI Canvas
    public Canvas canvas;
```

```csharp
//保存地图信息的 Text 类型的二维数组
private Text[,] mapText;

void Update()
{
    DetectPlayer();
    DetectInput();
    Judge();
    MovePlayer();
    ShowMap();
}
......
//显示地图信息
private void ShowMap()
{
    if(mapText == null)
    {
        mapText = new Text[9,9];  //新建二维数组，用来存储 Tile 信息
    }
    for(int i=0;i<9;i++)
    {
        for(int j=0;j<9;j++)
        {
            if(mapText[i,j] == null)
            {
                //实例化 Text
                mapText[i,j] = Instantiate(tileTypeText);
                //设置 Text 的父对象为 Canvas
                mapText[i,j].transform.parent = canvas.transform;
                //设置 Text 的位置
                mapText[i,j].transform.position = new Vector3(j*15, -i*15,0);
            }
            //将 Tile 信息赋予 Text
            mapText[i,j].text = ((int)map.GetTileType(new Vector2(j,-i),
                                Direction.Null)).ToString();
        }
    }
}
```

返回 Unity，将之前保存的 Text 预制体和场景中的 Canvas 赋值给 GameController 脚本，设置打印地图快照，如图 5-19 所示。运行游戏，打印地图快照信息，如图 5-20 所示。

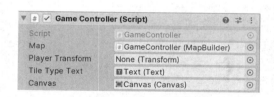

图 5-19 设置打印地图快照

图 5-20 地图快照信息

彩色图

5.4.8 角色移动与地图快照信息的实时更新

为了实现在角色移动时实时更新地图快照信息，需要在角色移动时改变地图信息数组中的值。MapBuilder 脚本中的代码如下：

```
public class MapBuilder : MonoBehaviour
{
......
//修改地图快照信息
public void SetTileType(Vector2 pos,GameController.Direction direction,TileType tiltType)
    {
        snapshotMap[-(int)pos.y*rowNumber+(int)pos.x (int)direction] = (int)tiltType;
    }
}
```

GameController 脚本中的代码如下：

```
public class GameController : MonoBehaviour
{
  ......
private void Judge()
{
 switch(map.GetTileType(playerPos,dir))
  {
   case MapBuilder.TileType.Null:
        //修改角色移动后位置的地图快照信息
        map.SetTileType(playerPos,dir,MapBuilder.TileType.Player);
        //修改角色原本坐标位置的地图快照信息
        //如果是普通地板
        if (map.GetTileType(playerPos,Direction.Null) ==
          MapBuilder.TileType.Player)
            map. SetTileType(playerPos,dir,MapBuilder.TileType.Null);
        //如果是目标点
        else
            map. SetTileType(playerPos,dir,MapBuilder.TileType.Target);
     canMovePlayer = true;
     break;
    ......
       }
   }
}
```

5.4.9 推动箱子的功能实现

（1）箱子的移动与角色的移动算法类似，只是除了刷新角色位置，还需要刷新箱子的位置。因此，移动箱子其实就是设定 dir 控制角色在数组中的移动距离。例如，向右移动，此时 dir＝1。将角色目前位置的数组值变为 0，下一个目标移动位置的数组值变为 2，下下个目标移动位置的数组值变为 3，刷新地图数组（游戏状态快照）。可以使用一个数组来保存地图中的箱子数据。MapBuilder 脚本中的代码如下：

```
public class MapBuilder : MonoBehaviour
{
  ......
```

```csharp
//保存当前场景中箱子的数组
private GameObject[] boxObjects;
……
private void BuildMap()
{
    for(int row = 0; row < rowNumber; row++)
    {
        for(int col = 0; col < colNumber;col++)
        {
            switch(snapshotMap[row * rowNumber + col])
            {
            ……
            case (int)TileType.Box:
                //在生成箱子时,把箱子保存在数组中
                boxObjects[row*9+col] = BuildTile(row,col,box);
                break;
            }
        }
    }
}

//获得箱子
public GameObject GetBox(Vector2 pos,GameController.Direction direction)
{
    return boxObjects[-(int)pos.y * rowNumber + (int)pos.x + (int)direction];
}
//设置箱子
public void SetBox(Vector2 pos, GameObject box)
{
    boxObjects[-(int)pos.y * rowNumber + (int)pos.x] = box;
}
```

GameController 脚本中的代码如下:

```csharp
public class GameController : MonoBehaviour
{
    ……
    //箱子是否可以移动
    private bool canMoveBox = false;
    ……
    private void Judge()
    {
        switch(map.GetTileType(playerPos,dir))
        {
        ……
            case MapBuilder.TileType.Box:
            switch(map.GetTileType(playerPos,(Direction)((int)dir*2)))
            {
                case MapBuilder.TileType.Null:
                    //将箱子的类型修改为角色的类型
                    map.SetTileType(playerPos,dir,MapBuilder.TileType.Player);
                    //箱子下一个移动的位置改为箱子的类型
                    map.SetTileType(playerPos,(Direction)((int)(dir*2)),
                        MapBuilder.TileType.Box);
```

```csharp
                canMovePlayer = true;          //角色可以移动
                canMoveBox = true;             //箱子可以移动
                if(map.GetTileType(playerPos,Direction.Null)==
                        MapBuilder.TileType.Player)
                    map.SetTileType(playerPos,dir,MapBuilder.TileType.Null);
                //如果是目标点
                else
                    map.SetTileType(playerPos,dir,MapBuilder.TileType.Target);
                break;
            case MapBuilder.TileType.TargetPoint:
                canMovePlayer = true;          //角色可以移动
                canMoveBox = true;             //箱子可以移动
                break;
        }
        break;
    ……
    default:
        canMoveBox = false;                    //箱子不能移动
        break;
    }
}

private void MoveBox()
{
    if(canMoveBox == true)
    {
        //获得当前的箱子对象
        //修改箱子数组中的内容
        //移动箱子
        //判断箱子是否已移动到目标点上
    }
}
```

（2）需要获取角色移动方向前方的箱子，同时更新地图快照信息，这与角色移动算法相似。MapBuilder 脚本中的代码如下：

```csharp
public class MapBuilder : MonoBehaviour
{
    ……
    private void InitMap()
    {
        //实例化箱子
        boxObjects = new GameObject[rowNumber*colNumber];
        ……
    }

    //加入位置信息
    public void SetBox(Vector2 pos,GameController.Direction direction,
                    GameObject box)
    {
        boxObjects[-(int)pos.y * rowNumber + (int)pos.x + (int)direction] = box;
    }
}
```

GameController 脚本中的代码如下：

```csharp
public class GameController : MonoBehaviour
{
    ……
    //普通箱子图片
    public Sprite normalBoxImage;
    //到达目标点的箱子图片
    public Sprite targetedBoxImage;
    ……
    private void MoveBox()
    {
        if(canMoveBox == true)
        {
            //获得当前的箱子对象
            GameObject currentBox = map.GetBox(playerPos,dir);
            //修改箱子数组中的内容
            map.SetBox(playerPos,dir,null);
            map.SetBox(playerPos,(Direction)((int)dir*2),currentBox);
            //移动箱子
            currentBox.transform.position = playerPos + nextPlayerPos*2;
            //判断箱子是否已移动到目标点上
            if(map.GetTileType(currentBox.transform.position,(int)Direction.Null) ==
                        MapBuilder.TileType.TargetWithBox)
            {
                currentBox.GetComponent<SpriteRenderer>().sprite =targetedBoxImage;
            }
            else{
                currentBox.GetComponent<SpriteRenderer>().sprite = normalBoxImage;
            }
        }
    }
}
```

（3）将普通箱子图片和到达目标点的箱子图片赋值给 GameController 脚本，如图 5-21 所示。

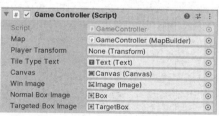

图 5-21　为 GameController 脚本赋值

（4）当把箱子推到目标点时，箱子颜色需要发生变化。当把箱子推到目标点时，可以更改目标点数据。GameController 脚本中的代码如下：

```csharp
public class GameController : MonoBehaviour
{
    ……
    private void Judge()
    {
        switch(map.GetTileType(playerPos,dir))
```

```
{
......
case MapBuilder.TileType.Box:
    switch(map.GetTileType(playerPos,(Direction)((int)dir*2)))
    {
    case MapBuilder.TileType.Null:
        ......
    case MapBuilder.TileType.TargetPoint:
        //将箱子的类型修改为角色的类型
        map.SetTileType(playerPos,dir,MapBuilder.TileType.Player);
        //将箱子下一个移动的位置改为目标点和箱子的类型
        map.SetTileType(playerPos,(Direction)((int)(dir*2)),
            MapBuilder.TileType.TargetWithBox);
        canMovePlayer = true;        //角色可以移动
        canMoveBox = true;           //箱子可以移动
        if(map.GetTileType(playerPos,Direction.Null) ==
          MapBuilder.TileType.Player)
             map.SetTileType(playerPos,dir,MapBuilder.TileType.Null);
        //如果是目标点
        else
           map.SetTileType(playerPos,dir,MapBuilder.TileType.Target);
        break;
    }
    break;
case MapBuilder.TileType.TargetWithBox:
    switch(map.GetTileType(playerPos,(Direction)((int)dir*2)))
    {
        case MapBuilder.TileType.Null:
            //角色的下一个位置的类型为目标点和角色的类型
            map.SetTileType(playerPos,dir,
                        MapBuilder.TileType. TargetWithPlayer);
            //将箱子下一个移动的位置改为箱子的类型
            map.SetTileType(playerPos,(Direction)((int)(dir*2)),
                        MapBuilder.TileType.TargetWithBox);
            canMovePlayer = true;        //角色可以移动
            canMoveBox = true;           //箱子可以移动
            if(map.GetTileType(playerPos,Direction.Null) ==
                MapBuilder.TileType.Player)
                  map.SetTileType(playerPos,dir,MapBuilder.TileType.Null);
            //如果是目标点
            else
                map.SetTileType(playerPos,dir,MapBuilder.TileType.Target);
            break;
    }
    break;
case MapBuilder.TileType.TargetPoint:
    //将目标点和箱子的类型修改为目标点和角色的类型
    map.SetTileType(playerPos,dir,MapBuilder.TileType. TargetWithPlayer);
    //将箱子下一个移动的位置改为目标点和箱子的类型
    map.SetTileType(playerPos,(Direction)((int)(dir*2)),
          MapBuilder.TileType.TargetWithBox);
```

```
            canMovePlayer = true;              //角色可以移动
            canMoveBox = true;                 //箱子可以移动
            if(map.GetTileType(playerPos,Direction.Null) ==
              MapBuilder.TileType.Player)
                    map.SetTileType(playerPos,dir,MapBuilder.TileType.Null);
            //如果是目标点
            else
                 map.SetTileType(playerPos,dir,MapBuilder.TileType.Target);
            break;
            ……
        }
    }
}
```

（5）修改图片的图层数。将 Target 的 Order in Layer 修改为 1，Box 的 Order in Layer 修改为 2，Player 的 Order in Layer 修改为 3。修改图片的图层数后的效果如图 5-22 所示。

图 5-22　修改图片的图层数后的效果

5.4.10　移动代码的重构优化

（1）为了使代码更加简洁，可以将重复使用的语句尽量提取到一个新的函数中，并在原位置对其进行调用。GameController 脚本中的代码如下：

```
public class GameController : MonoBehaviour
{
    ……
    //更改角色的状态
    private void ChangePlayerState()
    {
        if(map.GetTileType(playerPos,Direction.Null) == MapBuilder.TileType.Player)
        {
            map.SetTileType(playerPos,Direction.Null,MapBuilder.TileType.Null);
        }
        else{
            map.SetTileType(playerPos,Direction.Null,
                            MapBuilder.TileType.TargetPoint);
        }
    }

    //更改箱子的状态
    private void ChangeStateWithBox(Vector2 playerPos,Direction dir,
```

```csharp
                            MapBuilder.TileType tileType01,
                            MapBuilder.TileType tileType02)
{
    map.SetTileType(playerPos,dir,tileType01);
    map.SetTileType(playerPos,(Direction)((int)dir*2),tileType02);
    canMovePlayer = true;
    canMoveBox = true;
}

private void Judge(){
    switch(map.GetTileType(playerPos,dir))
    {
    case MapBuilder.TileType.Null:
        map.SetTileType(playerPos,dir,MapBuilder.TileType.Player);
        ChangePlayerState();    //更改角色的状态
        canMovePlayer = true;
        break;
    case MapBuilder.TileType.TargetPoint:
        map.SetTileType(playerPos,dir,MapBuilder.TileType.TargetWithPlayer);
        ChangePlayerState();            //更改角色的状态
        canMovePlayer = true;
        break;
    case MapBuilder.TileType.Box:
        switch(map.GetTileType(playerPos,(Direction)((int)dir*2)))
        {
        case MapBuilder.TileType.Null:
            //更改箱子的状态
            ChangeStateWithBox(playerPos,dir, MapBuilder.TileType.Player,
                        MapBuilder.TileType.Box);
            ChangePlayerState();      //更改角色的状态
            break;
        case MapBuilder.TileType.TargetPoint:
            //更改箱子的状态
            ChangeStateWithBox(playerPos,dir, MapBuilder.TileType.Player,
                        MapBuilder.TileType.TargetWithBox);
            ChangePlayerState();      //更改角色的状态
            break;
        }
        break;
    case MapBuilder.TileType.TargetWithBox:
        switch(map.GetTileType(playerPos,(Direction)((int)dir*2)))
        {
            case MapBuilder.TileType.Null:
                //更改箱子的状态
                ChangeStateWithBox(playerPos,dir,  MapBuilder.TileType.TargetWithPlayer,
                            MapBuilder.TileType.Box);
                ChangePlayerState();        //更改角色的状态
                break;
            case MapBuilder.TileType.TargetPoint:
                //更改箱子的状态
                ChangeStateWithBox(playerPos,dir,
```

```
                                    MapBuilder.TileType.TargetWithPlayer,
                                    MapBuilder.TileType.TargetWithBox);
            ChangePlayerState();        //更改角色的状态
            break;
        }
        break;
        ……
    }
}
```

（2）将之前设置好的动画状态机在代码中进行调用，从而改变角色的状态。GameController 脚本中的代码如下：

```
public class GameController : MonoBehaviour
{
    ……
    //动画组件
    private Animator anim;
    ……

    private void DetectPlayer()
    {
        if(playerTransform == null)
        {
            playerTransform = GameObject.Find("Player").GetComponent<Transform>();
            //获得角色的动画组件
            anim = playerTransform.gameObject.GetComponent<Animator>();
        }
        playerPos = playerTransform.position;
    }

    private void MovePlayer()
    {
        switch(dir)
        {
        case Direction.Up:
            nextPlayerPos = new Vector3(0,1);
            anim.Play("BackAnim");   //向后的动画
            break;
        case Direction.Down:
            anim.Play("FrontAnim"); //向前的动画
            nextPlayerPos = new Vector3(0,-1);
            break;
        case Direction.Left:
            anim.Play("LeftAnim");   //向左的动画
            nextPlayerPos = new Vector3(-1,0);
            break;
        case Direction.Right:
            anim.Play("RightAnim"); //向右的动画
            nextPlayerPos = new Vector3(1,0);
            break;
        case Direction.Null:
            nextPlayerPos = new Vector3(0,0);
```

```
            break;
        }
        ......
    }
}
```

5.4.11 游戏胜利条件判断

（1）添加游戏胜利图片。创建一个计数变量，并设置为当前关卡中目标点的数量，当箱子被移动到目标点时，该变量的值-1；当箱子被移出目标点时，该变量的值+1；当该变量的值为 0 时，游戏结束。在场景中创建一张游戏胜利图片，先将该图片拖入 Source Image，再将该图片的 Active 关闭，如图 5-23 所示。

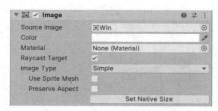

图 5-23　添加游戏胜利图片

（2）在代码中使用一个计数变量记录剩余目标点的数量，当计数变量的值为 0 时显示游戏胜利图片。MapBuilder 脚本中的代码如下：

```
public class MapBuilder : MonoBehaviour
{
    ......
    //剩余目标点的数量
    private int targetCount = 0;
    ......
    private void BuildMap()
    {
        for(int row = 0; row < rowNumber; row++)
        {
            for(int col = 0; col < colNumber;col++)
            {
                switch(snapshotMap[row * rowNumber + col])
                {
                    ......
                    case (int)TileType.TargetPoint:
                        BuildTile(row,col,targetPoint);
                        targetCount++;   //获取地图加载时目标点的数量
                        break;
                    ......
                }
            }
        }
    }
    ......
    //增加目标点的数量
    public void AddTargetCount() {targetCount++;}
```

```csharp
    //减少目标点的数量
    public void ReduceTargetCount()
    {
        if(targetCount > 0)
        {
            targetCount--;
        }
    }
    //获得目标点的数量
    public int GetTargetCount()
    {
        return targetCount;
    }
}
```

GameController 脚本中的代码如下：

```csharp
public class GameController : MonoBehaviour
{
    ......
    //游戏胜利图片
    public Image winImage;

    void Update()
    {
        ......
        GameOver();
    }

    //当计数变量的值为 0 时显示游戏胜利图片
    private void GameOver()
    {
        if(map.GetTargetCount() == 0)
        {
            winImage.gameObject.SetActive(true);
        }
    }

    private void Judge()
    {
        switch(map.GetTileType(playerPos,dir))
        {
        ......
        case MapBuilder.TileType.Box:
            switch(map.GetTileType(playerPos,(Direction)((int)dir*2)))
            {
            ......
            case MapBuilder.TileType.TargetPoint:
                ChangeStateWithBox(playerPos,dir, MapBuilder.TileType.Player,
                    MapBuilder.TileType.TargetWithBox);
                ChangePlayerState();
                    map.ReduceTargetCount();      //减少目标点的数量
                    break;
            }
            break;
```

```
                case MapBuilder.TileType.TargetWithBox:
                    switch(map.GetTileType(playerPos,(Direction)((int)dir*2)))
                    {
                    case MapBuilder.TileType.Null:
                        ChangeStateWithBox(playerPos,dir, MapBuilder.TileType.Target,
                                        WithPlayer,MapBuilder.TileType.Box);
                        ChangePlayerState();
                        map.AddTargetCount();              //增加目标点的数量
                        break;
                        ……
                    }
                    ……
            }
        }
    }
```

5.5 本章小结

本章主要介绍了《推箱子》游戏的简单实现，其中重点在于地图中各种元素的抽象表示方法、角色与箱子在移动时涉及的各种情况的判定，以及获胜的各种条件的判定等。

5.6 练习题

1. 本章讲述的《推箱子》游戏的核心实现涉及大量的条件判断语句，用来判定角色和箱子在各种情况下是否可以移动。现换一个角度重新考虑：将每个预制体（包括角色、箱子、墙和目标点等元素）均看作有一定自我意识的独立个体，即在每个预制体上分别设置各自独立的实现脚本，描述其在遇到外力作用时应该如何做出移动反应。请问此种实现方式是否可行？简要列出这种实现方式相比原实现方式的优点和缺点。如果读者有其他的实现思路，亦可简要描述。

2. 本章讲述的《推箱子》游戏的获胜条件为"在所有目标点均放入箱子"，现若将其改为"在每个目标点上均放入对应的箱子"（为箱子与目标点分别编号，要对号入座才可以），应该如何实现呢？请从数据结构的角度简述实现思路。

3. （选做）试思考在设计《推箱子》游戏的地图时应满足何种条件才能保证一定有解（至少存在一套合理步骤，使所有箱子均可到达指定目标点），而不是在中途就宣告失败。

第 6 章

俄罗斯方块

6.1 游戏简介

《俄罗斯方块》是游戏制作人帕基特诺夫制作的一款经典的休闲游戏。1984 年 6 月，在苏联科学院计算机中心工作的帕基特诺夫利用空闲时间编写了一个游戏程序，用来测试当时一种计算机的性能。帕基特诺夫爱玩拼图游戏，从拼图游戏中得到了灵感，从而设计出了该游戏。1989 年的《俄罗斯方块》游戏的画面，如图 6-1 所示。

图 6-1　1989 年的《俄罗斯方块》游戏的画面

6.2 游戏规则

（1）游戏场景。有一块用于摆放方块的平面虚拟场地，其标准大小如下：行宽为 10，列高为 20，以一个方块为一个单位。

（2）有一组由 4 个方块组成的规则图形（中文通称为方块），共有 7 种，分别以 S、Z、L、J、I、O、T 这 7 个字母的形状来命名，如图 6-2 所示。

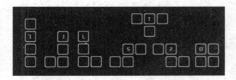

图 6-2　《俄罗斯方块》游戏的 7 种图形

（3）玩家可以以 90°为单位旋转方块，以格子为单位向左或向右移动方块，或者让方块加速下落。方块移到区域最下方或着地到其他方块上无法移动时，就固定在该处，此时新的方块会出现在区域上方开始落下。一般来说，游戏会提示下一个要落下的方块。

（4）当区域中某行的格子全部由方块填满时，该行的方块会消失并成为玩家的得分。删除的行越多，得分越高。当方块堆到区域最上方而无法消除行时，该游戏结束。

（5）常见的游戏模式有以下几种。
- 经典马拉松模式：如果选择该模式，那么游戏无计时，一直到方块堆到最上方并且无法消除时结束游戏。
- 竞速模式：消除指定的行数，用时最短者获胜。
- 定时模式：在一定的时间内，得分最高者获胜。
- 重力模式：传统版本的《俄罗斯方块》游戏将堆栈的块向下移动一段距离，正好等于它们之下清除的行的高度。与重力定律相反，块可以悬空在间隙之上，如图 6-3 所示。实现使用洪水填充的不同算法将游戏区域分割成连接的区域将使每个区域并行落下，直到它接触到底部的区域。这开启了额外的"连锁反应"策略，涉及块级联以填补额外的线路，这可能被认为是更有价值的清除，如图 6-4 所示。

图 6-3　块悬空在间隙之上　　　　　图 6-4　重力模式下的"连锁反应"

本章实现的是《俄罗斯方块》游戏的经典马拉松模式。

（6）降落方式包括以下两种。
- 硬降：方块立即下落到最下方并锁定。
- 软降：方块加速下落。

（7）旋转踢墙：一个方块在即使方块的边界已经碰到墙体的情况下，如果旋转后有足够空间则能继续旋转。在 NES 版本中，如果一个 Z 形方块竖立着靠着左侧的墙体，方块即使有足够的空间也不能旋转这个方块，感觉像是锁定了一样。在这种情况下，玩家必须在旋转之前把方块向右移动一格，但这会失去宝贵的时间。最早的踢墙出现在 Atari 的街机版《俄罗斯方块》游戏中。

6.3　游戏开发核心思路

6.3.1　初始化地图

使用二维数组生成当前地图状态，记录游戏场景状态、绘制背景，并且以地图的数值为依据进行边界判断。每次移动方块都要更新地图。共有 4 种地图块，分别为墙体、空格、固定方块（方块不可以再操作时）和游戏中的方块（方块的形状）。使用二维数组表达游戏场景的示例如图 6-5 所示，使用不同 ID 表达方块不同形状的示例如图 6-6 所示。

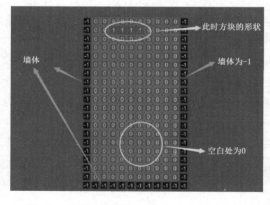
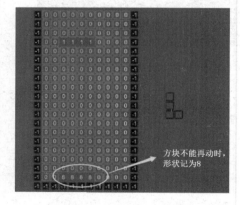

图 6-5　使用二维数组表达游戏场景的示例　　　图 6-6　使用不同 ID 表达方块不同形状的示例

6.3.2 方块类（Block）的编写

初始化方块的形状可以使用二维数组，通过数字存储方块的 4 种变换形状。用一个 4×16 的二维数组记录不同类型方块在不同朝向时的形状，4 表示方块的 4 种旋转朝向，16 用于记录方块的形状。需要根据方块类型来决定方块的变换形状。

存储原则包括以下几点。
- 方块由小方块组成，数字代表一个小的组成方块。
- 用数组抽象表示方块的形状，不为 0 的数字表示有小方块，0 表示没有小方块。
- 数字与当前方块类型有关，不同类型方块的数值不同。
- 二维数组中每个花括号里的数组都表示一种形状的方块。

除此之外，还需要编写获取操作，获取方块形状和方块当前位置，以及编写变换操作，以便能够旋转、重置和移动（包括向下、向左或向右）方块。

6.3.3 初始化方块——方块在地图中的随机生成

先把 7 种方块放在一个一维数组中，再用随机遍历的方法在场景中生成两个方块，一个用于当前方块的控制，另一个用于下一个方块的控制。

6.3.4 移动和旋转方块——修改位置坐标

在移动方块时采用先移动后判断的方式，若满足任意一个不可移动的条件，则撤销移动，否则判断为成功移动，并更新地图（旋转同理）。移动方块时坐标的修改如下。
- 向左移动：方块位置 $x-1$。
- 向右移动：方块位置 $x+1$。
- 旋转：在方块形状数组中取下一种形状。

判断流程图如图 6-7 所示。

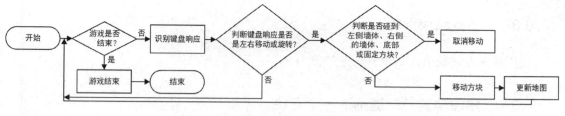

图 6-7 判断流程图

6.3.5 移动和旋转方块——边界判断

在移动或旋转方块后，遍历方块类中的每个元素，如果该元素不为 0，那么需要进行以下判断。
- 碰到左侧墙体或右侧墙体。

- 碰到底部。
- 碰到其他方块。

6.3.6 移动方块——向下移动

碰撞判断包括以下几点。
- 碰到底部：方块所在位置的左侧、右侧或下方的小方块的数值< 0。
- 碰到其他方块：地图信息为 8。
- 在地图中更新固定方块信息：将当前移动的方块形状变更为 8。
- 更新地图：变更地图二维数组中的数值。

6.3.7 消除满行方块

消除满行方块流程图如图 6-8 所示。

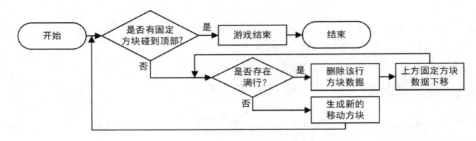

图 6-8 消除满行方块流程图

是否碰到顶部的碰撞条件为地图最上面一行存在数值 8。
- 满行判断：一行一行地遍历地图（除去边界），若扫描到不是 8 的数字，则跳出此行，遍历下一行。若地图上某一行的数值全为 8，则为满行。
- 满行消除：从满行的该行往上的所有行，固定方块数据全部下移。

6.3.8 游戏结束判断

遍历最高行的数值，如果有一个数值为 8，则表示游戏结束。

6.3.9 附加功能——提示下一个方块

在场景中的右侧显示下一个要下落的方块类型。使用一个 4×4 的二维数组来显示方块提示。

6.3.10 游戏流程图

游戏流程图如图 6-9 所示。

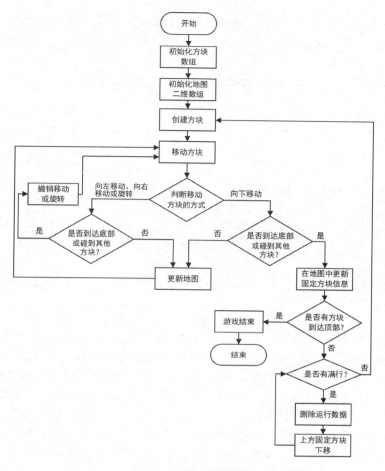

图 6-9　游戏流程图

6.4　游戏实现

6.4.1　资源的导入

（1）新建工程。新建一个名称为 Tetris 的工程，选择 2D 选项，如图 6-10 所示。

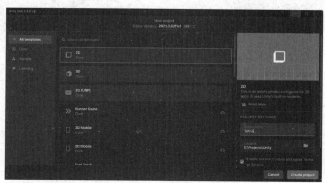

图 6-10　新建工程

（2）导入资源。需要创建不同的文件夹，这是因为要将不同的资源放到不同的文件夹中。将下载的图片资源放到 Sprites 文件夹中。将两张图片的 Inspector 窗口中的 Pixels Per Unit 设置为 64，如图 6-11 所示。

（3）设置摄像机。将 Main Camera 的 Position 的 X 和 Y 都设置为 10，Size 设置为 11，如图 6-12 所示。

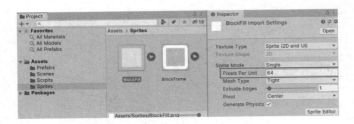

图 6-11　导入资源　　　　　　　　　　　　　图 6-12　设置摄像机

（4）制作地图初始方块的预制体。首先，制作墙体。将 Sprites 文件夹下的 BlockFrame 图片和 BlockFill 图片拖动到 Scene 窗口中，并且使 BlockFill 图片成为 BlockFrame 图片的子物体。将 BlockFill 的 Transform 重置后，先将 BlockFrame 图片的 Sprite Renderer 组件的 Color 改为灰色，并稍微降低其 alpha 值，再将 BlockFill 图片的 Sprite Renderer 组件的 Color 改为黑色，来代表墙体，稍微降低其 alpha 值，如图 6-13 所示。在制作完成后，将 BlockFrame 图片拖到 Prefabs 文件夹下，使其成为一个预制体，并重命名为 Wall，删除 Scene 窗口中的原物体。接着，制作空白位置。再一次将 BlockFrame 图片拖入 Scene 窗口中，把 BlockFrame 图片的颜色改为灰色，并降低其 alpha 值，使其变得稍微透明一些，如图 6-14 所示。在制作完成后，将 BlockFrame 图片拖到 Prefabs 文件夹下，使其成为一个预制体，并将其命名为 BlockFrame，删除 Scene 窗口中的原物体，如图 6-15 所示。

图 6-13　制作墙体　　　　　　　　　　　　图 6-14　制作空白位置

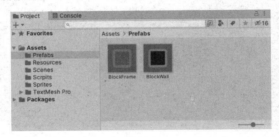

图 6-15　地图初始方块的预制体

6.4.2 地图的初始化与显示

（1）编写地图初始化脚本。打开 Scripts 文件夹，新建一个脚本并将其命名为 Map。在 InitMap()函数中用两个 for 循环来初始化地图快照数组和地图方块的 Sprites 数组。在 for 循环中编写一个判断，若遇到应是墙体的列、行时，使用 Instantiate()函数实例化墙体，否则使用 Instantiate()函数实例化空格；更改两个数组的信息；在 Start()函数中调用 InitMap()函数。Map 脚本中的代码如下：

```csharp
using System.Collections;
using System.Collections.Generic;
using UnityEngine;

public class Map : MonoBehaviour
{
    public static int mapRow = 20;
    public static int mapCol = 12;
    public GameObject blockFrame;                       //空格
    public GameObject blockWall;                        //墙体
    private GameObject[,] backgroundObjs;               //地图方块数组
    private int[,] mapSnapshot;                         //地图快照
    //Start is called before the first frame update
    void Start()
    {
        InitMap();
    }

    //初始化地图
    void InitMap()
    {
        backgroundObjs = new GameObject[mapRow, mapCol];    //绘制背景
        mapSnapshot = new int[mapRow, mapCol];              //记录游戏场景状态
        for (int row = 0; row < mapRow; row++)
        {
            for (int col = 0; col < mapCol; col++)
            {
                //左侧、底部和右侧为墙体
                if (col == 0 || col == mapCol - 1 || row == 0)
                {
                    //墙体
                    mapSnapshot[row, col] = -1;
                    backgroundObjs[row, col] = Instantiate(blockWall,
                                        new Vector2(col, row),
                                        Quaternion.identity);
                    backgroundObjs[row, col].name = "Wall" + row + "_" + col;
                }
                else
                {
                    //空格
                    mapSnapshot[row, col] = 0;
                    backgroundObjs[row, col] = Instantiate(blockFrame,
                                        new Vector2(col, row),
                                        Quaternion.identity);
```

```
                backgroundObjs[row, col].name = "Frame" + row + "_" + col;
            }
            //把所有的背景对象都放在一个父物体中,保持场景的整洁
            backgroundObjs[row, col].transform.SetParent(transform);
        }
    }
}
```

(2)在 Hierarchy 窗口的空白处单击鼠标右键,在弹出的快捷菜单中选择 Create Empty 命令,并将新建的空物体重命名为 Map,单击 Reset 按钮重置 Map 物体的位置。将 Map 脚本挂载到 Map 物体上。将 BlockFrame 预制体赋值给 Map 脚本中的 BlockFrame,将 Wall 预制体赋值给 Map 脚本中的 BlockWall,如图 6-16 所示。

(3)运行游戏并查看效果,生成的初始地图如图 6-17 所示。

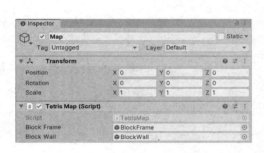

图 6-16　挂载脚本并赋值

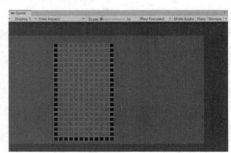

图 6-17　生成的初始地图

6.4.3　地图快照信息的显示

可以使用 TextMeshPro 来显示地图快照的文字信息。

(1)在 Unity 界面中,选择 GameObject→3D Object→Text-TextMeshPro 命令,新建一个 TextMeshPro,按照如图 6-18 所示修改其属性。

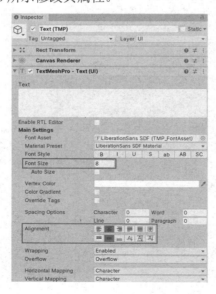

图 6-18　新建 3D Text

在第一次添加 TextMeshPro 时，需要先导入必要的文件（TMP），如图 6-19 所示。

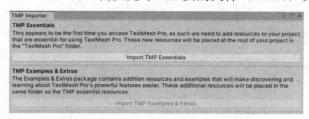

图 6-19　导入必要的文件（TMP）

当 TMP 报错时，需要为 TextMeshPro 设置 Material 和 Font Asset，外部导入的字体需要制作成 Font Asset 才可以使用。制作 Font Asset 的步骤如图 6-20 所示。

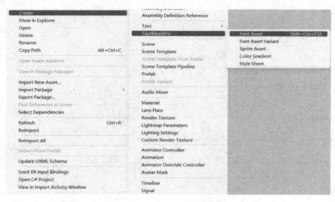

图 6-20　制作 Font Asset 的步骤

（2）将该 Text (TMP)重命名为 InfoText，并将其拖到 Prefabs 文件夹中，使其成为一个预制体；删除场景中的原物体。

（3）打印地图快照信息。需要编写一个 ShowMapInfo()函数来打印地图快照信息，并在 Start()函数中调用一次该函数。代码如下：

```csharp
using System.Collections;
using System.Collections.Generic;
using UnityEngine;

public class Map : MonoBehaviour
{
    public static int mapRow = 20;
    public static int mapCol = 12;
    public GameObject blockFrame;              //空格
    public GameObject blockWall;               //墙体
    private GameObject[,] backgroundObjs;      //地图的方块数组
    private int[,] mapSnapshot;                //地图快照
    private GameObject textRoot;               //用于存储显示地图标记信息的父物体
    public TextMeshPro mapText;
    private TextMeshPro [,] mapInfoTexts;      //显示地图的标记信息
    // Start is called before the first frame update
    void Start()
    {
        InitMap();
        ShowMapInfo();
```

```
}

//初始化地图
void InitMap(){……}

//显示地图信息,用于测试
public void ShowMapInfo()
{
    if (mapInfoTexts == null)
    {
        mapInfoTexts = new TextMeshPro [mapRow, mapCol];
        textRoot = new GameObject("TextRoot");
    }

    //遍历地图快照信息
    for (int row = 0; row < mapRow; row++)
    {
        for (int col = 0; col < mapCol; col++)
        {
            if (mapInfoTexts[row, col] == null)
            {
                mapInfoTexts[row, col] = Instantiate(mapText,
                                                    new Vector3(col, row, 0),
                                                    Quaternion.identity);
                mapInfoTexts[row, col].transform.SetParent(textRoot.transform);
            }
            //获得第 row 行第 col 列的快照信息
            mapInfoTexts[row, col].text = mapSnapshot[row, col].ToString();
        }
    }
}
```

（4）拖入预制体。在保存代码后，返回 Unity 界面，将 InfoText 预制体赋值给 Map 脚本中的 Map Text，如图 6-21 所示。

在完成上述步骤后，运行游戏，地图快照信息便出现可视化效果，如图 6-22 所示。

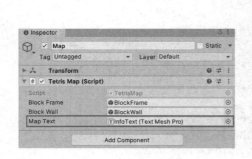

图 6-21　拖入预制体

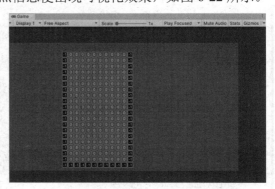

图 6-22　可视化效果

6.4.4　方块的初始化

接下来编写方块类（Block）。

（1）打开 Scripts 文件夹，新建脚本并将其命名为 Block。使用枚举来表示方块的类型。可以使用二维数组通过数字存储方块的 4 种变换形状。例如，使用一个 4×16 的二维数组，4 表示方块 4 种不同的旋转朝向，16 用于记录方块的形状。将这个二维数组称为方块形状数组，方块类型决定了方块的变换形状。具体的存储原则如下：方块由小方块组成，数字代表一个小的组成方块；用数组抽象表示方块的形状，不为 0 的数字表示有小方块，0 表示没有小方块；数字与当前方块形状有关，不同类型方块的数字不同；二维数组中每个花括号里的数组都表示一种形状的方块。

Block 脚本中的代码如下：

```csharp
using System.Collections;
using System.Collections.Generic;
using UnityEngine;

public class Block : MonoBehaviour
{
    public int[,] shape { private set; get; }         //方块的形状
    public enum BlockType { I, J, L, O, S, Z, T }     //枚举类型，代表方块的形状
    public BlockType type;
    public Color blockColor;                          //区分每个方块的颜色
    private int curState = 0;                         //当前的形状

    public void InitShape()
    {
        switch (type)
        {
            case BlockType.I:
                shape = new int[4, 16] {
                    {0,0,0,0,
                     1,1,1,1,
                     0,0,0,0,
                     0,0,0,0},          //横向

                    {0,1,0,0,
                     0,1,0,0,
                     0,1,0,0,
                     0,1,0,0},          //纵向

                    {0,0,0,0,
                     1,1,1,1,
                     0,0,0,0,
                     0,0,0,0},          //横向

                    {0,1,0,0,
                     0,1,0,0,
                     0,1,0,0,
                     0,1,0,0}           //纵向
                };
                break;
            case BlockType.J:
                shape = new int[4, 16] {
                    {0,2,0,0,
                     0,2,0,0,
```

```
            2,2,0,0,
            0,0,0,0}, //J

           {2,2,2,0,
            0,0,2,0,
            0,0,0,0,
            0,0,0,0},

           {2,2,0,0,
            2,0,0,0,
            2,0,0,0,
            0,0,0,0},

           {2,0,0,0,
            2,2,2,0,
            0,0,0,0,
            0,0,0,0}
        };
        break;
case BlockType.L:
        shape = new int[4, 16] {
           {3,0,0,0,
            3,0,0,0,
            3,3,0,0,
            0,0,0,0}, //L

           {0,0,3,0,
            3,3,3,0,
            0,0,0,0,
            0,0,0,0},

           {3,3,0,0,
            0,3,0,0,
            0,3,0,0,
            0,0,0,0},

           {3,3,3,0,
            3,0,0,0,
            0,0,0,0,
            0,0,0,0}
        };
        break;
case BlockType.O:
        shape = new int[4, 16] {
           {4,4,0,0,
            4,4,0,0,
            0,0,0,0,
            0,0,0,0}, //O

           {4,4,0,0,
            4,4,0,0,
            0,0,0,0,
            0,0,0,0},

           {4,4,0,0,
```

```
            4,4,0,0,
            0,0,0,0,
            0,0,0,0},

           {4,4,0,0,
            4,4,0,0,
            0,0,0,0,
            0,0,0,0}
        };
        break;
   case BlockType.S:
        shape = new int[4, 16] {
           {0,5,5,0,
            5,5,0,0,
            0,0,0,0,
            0,0,0,0},  //S

           {5,0,0,0,
            5,5,0,0,
            0,5,0,0,
            0,0,0,0},

           {0,5,5,0,
            5,5,0,0,
            0,0,0,0,
            0,0,0,0},

           {5,0,0,0,
            5,5,0,0,
            0,5,0,0,
            0,0,0,0}
        };
        break;
   case BlockType.Z:
        shape = new int[4, 16] {
           {6,6,0,0,
            0,6,6,0,
            0,0,0,0,
            0,0,0,0},  //Z

           {0,6,0,0,
            6,6,0,0,
            6,0,0,0,
            0,0,0,0},

           {6,6,0,0,
            0,6,6,0,
            0,0,0,0,
            0,0,0,0},

           {0,6,0,0,
            6,6,0,0,
            6,0,0,0,
            0,0,0,0}
        };
```

```
            break;
        case BlockType.T:
            shape = new int[4, 16] {
                {0,7,0,0,
                 7,7,7,0,
                 0,0,0,0,
                 0,0,0,0},

                {0,7,0,0,
                 7,7,0,0,
                 0,7,0,0,
                 0,0,0,0},

                {7,7,7,0,
                 0,7,0,0,
                 0,0,0,0,
                 0,0,0,0},

                {0,7,0,0,
                 0,7,7,0,
                 0,7,0,0,
                 0,0,0,0}
            };
            break;
    }
}
```

（2）返回 Unity 界面，创建一个空的游戏对象并将其命名为 Block_I，将该对象挂载到 Block 脚本上。将 Type 设置为 I，BlockColor 设置为红色。将 Block_I 拖到 Prefabs 文件夹中，并删除场景中的原物体。

以此类推，将其他几种方块类型的预制体都制作好，需要为每种方块设置对应的 Type 和独有的颜色（颜色可以自行设置，只要能区分彼此即可）。图 6-23 所示为制作完成的 7 种方块的预制体。

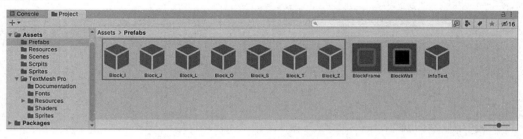

图 6-23 制作完成的 7 种方块的预制体

6.4.5 方块的随机生成

方块的随机生成，即随机地实例化一个方块预制体。当第一次生成方块时，需要特殊考虑，生成两个方块，分别为 curBlock（当前方块）和 nextBlock（下一个方块）。如果不是第一次生成方块，则需要先销毁 curBlock，再将 nextBlock 赋给 curBlock，最后 nextBlock 随机取新的方块。

（1）打开 Scripts 文件夹，新建脚本并将其命名为 GameController。在 GameController 脚本中编写 CreateBlock()函数，并在 Start()函数中调用一次该函数。运用 Debug.Log 进行测试，观察方块是否是随机生成的。GameController 脚本中的代码如下：

```csharp
using System.Collections;
using System.Collections.Generic;
using UnityEngine;

public class GameController : MonoBehaviour
{
    public Block[] blocks;                   //存储 7 种方块类型
    private Block curBlock;
    private Block nextBlock;
    private bool isFirst = true;             //是否第一次生成方块

    //Start is called before the first frame update
    void Start()
    {
        CreateBlock();
    }

    //生成方块
    void CreateBlock()
    {
        //如果游戏刚开始
        if (isFirst == true)
        {
            //随机地实例化一个方块预制体
            int curBlockIndex = Random.Range(0, blocks.Length);
            curBlock = Instantiate(blocks[curBlockIndex]);
            isFirst = false;

            int nextBlockIndex = Random.Range(0, blocks.Length);
            nextBlock = Instantiate(blocks[nextBlockIndex]);

            //在 Console 窗口中输出 curBlock 和 nextBlock 的名称
            Debug.Log(curBlock.name + " " + nextBlock.name);
        }
        else
        {
            Destroy(curBlock.gameObject);
            curBlock = nextBlock;
            nextBlock = Instantiate(blocks[Random.Range(0, blocks.Length)]);
        }
    }
}
```

（2）GameController 脚本的挂载与赋值。返回 Unity 界面，创建一个空的游戏对象并将其命名为 GameController，将该对象挂载到 GameController 脚本上。因为共有 7 种方块类型，所以将 Blocks 的 Size 设置为 7，并把方块预制体按照顺序拖到对应位置进行赋值，如图 6-24 所示。

（3）方块的随机生成。运行游戏，在 Console 窗口中可以看到，每次开始运行游戏，方块的生成都是随机的，如图 6-25 所示。

图 6-24 GameController 脚本的挂载与赋值

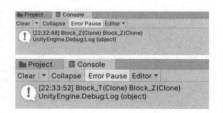

图 6-25 方块的随机生成

6.4.6 方块的移动和旋转

当方块移动时，若向左移动则方块位置 $x-1$，若向右移动则方块位置 $x+1$；当方块旋转时，就是在方块形状数组中取下一个形状。

（1）根据上述移动和旋转的原理，在 Block 类中编写移动和旋转方块的代码。Block 脚本中的代码如下：

```csharp
using System.Collections;
using System.Collections.Generic;
using UnityEngine;

public class Block : MonoBehaviour
{
    public int[,] shape { private set; get; }       //方块的形状
    public enum BlockType { I, J, L, O, S, Z, T }   //枚举类型，代表方块的形状
    public BlockType type;
    public Color blockColor;                         //区分每个方块的颜色
    private int curState = 0;                        //当前的形状

    private Vector2 curPos = new Vector2(3, Map.mapRow - 1);  //当前方块的位置
    private void Awake()
    {
        InitShape();
    }

    public void InitShape() {……}

    public Vector2 GetCurPos()
    {
        return curPos;
    }

    //向左移动
    public Vector2 MoveLeft()
    {
        curPos.x--;
```

```
        return curPos;
    }

    //向右移动
    public Vector2 MoveRight()
    {
        curPos.x++;
        return curPos;
    }

    //向下移动
    public Vector2 MoveDown()
    {
        curPos.y--;
        return curPos;
    }

    //向上移动
    public Vector2 MoveUp()
    {
        curPos.y++;
        return curPos;
    }

    //旋转方块
    public Vector2 RotateBlock()
    {
        curState = (curState + 1) % 4;
        return curPos;
    }

    //回撤旋转
    public Vector2 InverseRotateBlock()
    {
        curState = (curState + 3) % 4;
        return curPos;
    }

    public int GetBlockRotateState()
    {
        return curState;
    }
}
```

（2）返回 Map 脚本，先将地图快照清零的 ClearMap()函数和 SetMapItem()函数编写好，其中的代码暂时不编写。Map 脚本中的代码如下：

```
using System.Collections;
using System.Collections.Generic;
using UnityEngine;

public class Map : MonoBehaviour
{
    ......
```

```
    //刷新地图
    public void ClearMap()
    {
    }
    public void SetMapItem()
    {
    }
}
```

(3)返回 GameController 脚本,根据前面提到的移动的逻辑,先进行移动,再进行是否碰到墙体、其他方块的判断,如果判定为碰到了,则回到原来的位置,否则刷新地图,更新地图快照信息。按照这个逻辑,下面以向左移动为例编写大体的框架。GameController 脚本中的代码如下:

```
using System.Collections;
using System.Collections.Generic;
using UnityEngine;

public class GameController : MonoBehaviour
{
    public Block[] blocks;              //存储7种方块类型
    private Block curBlock;
    private Block nextBlock;
    private bool isFirst = true;        //是否第一次生成方块
    public Map map;

    //Start is called before the first frame update
    void Start()
    {
        CreateBlock();
    }

    void Update()
    {
        MoveBlock();
    }

    //生成方块
    void CreateBlock() {……}

    //移动和旋转方块
    void MoveBlock()
    {
        if (curBlock != null)
        {
            if (Input.GetKeyDown(KeyCode.LeftArrow))
            {
                //方块向左移动一个单位
                //MoveCurBlockLeft();
            }
            else if (Input.GetKeyDown(KeyCode.RightArrow))
            {
                //方块向右移动一个单位
```

```csharp
                //MoveCurBlockRight();
            }
            else if (Input.GetKeyDown(KeyCode.DownArrow))
            {
                //方块向下移动一个单位
                //MoveCurBlockDown();
            }
            else if (Input.GetKeyDown(KeyCode.UpArrow))
            {
                //方块旋转
                //RotateCurBlock();
            }
        }
    }

    void MoveCurBlockLeft()
    {
        //向左移动
        curBlock.MoveLeft();
        //判断是否碰到墙体或其他方块
        if (!CanMoveBlock(curBlock.GetCurPos()))
        {
            //如果不能向左移动,则回到原来的位置
            curBlock.MoveRight();
            return;
        }
        //刷新地图
        map.ClearMap();
        SetBlockToMap(curBlock.GetCurPos());
    }

    void MoveCurBlockRight()
    {

    }

    void MoveCurBlockDown()
    {

    }

    //判断方块是否可以移动
    bool CanMoveBlock(Vector2 curPos)
    {
        return false;
    }

    void SetBlockToMap(Vector2 curPos)
    {

    }
}
```

6.4.7 方块的移动、旋转与地图的刷新

本节主要介绍方块的移动监测和方块的旋转，以及地图的刷新。

（1）在 Map 脚本中对 ClearMap()函数和 SetMapItem()函数进行补充，代码如下：

```
using System.Collections;
using System.Collections.Generic;
using UnityEngine;

public class Map : MonoBehaviour
{
    //……

    private void Awake()
    {
        InitMap();
    }
    //Start is called before the first frame update
    void Start()
    {
        ShowMapInfo();
    }

    void Update()
    {
        ShowMapInfo();
    }

    //初始化地图
    void InitMap() {……}

    //显示地图信息，用于测试
    public void ShowMapInfo() {……}

    //刷新地图
    public void ClearMap()
    {
        for (int row = 0; row < mapRow; row++)
        {
            for (int col = 0; col < mapCol; col++)
            {
                if (mapSnapshot[row, col] != -1 && mapSnapshot[row, col] != 8)
                {
                    mapSnapshot[row, col] = 0;
                    backgroundObjs[row, col]
                      .GetComponentInChildren<SpriteRenderer>()
                      .color = new Color(1, 1, 1, 0.2f);
                }
            }
        }
    }

    public void SetMapItem(int row, int col, int flag, Color blockColor)
```

```
    {
        if (mapSnapshot == null)
        {
            return;
        }
        mapSnapshot[row, col] = flag;
        backgroundObjs[row, col].GetComponentInChildren<SpriteRenderer>()
            .color = blockColor;
    }
}
```

（2）在 GameController 脚本中，首先，根据已经编写好的 MoveCurBlockLeft()函数的框架，补全 MoveCurBlockRight()函数、MoveCurBlockDown()函数和 RotateCurBlock()函数的框架；然后，编写 SetBlockToMap()函数，把方块的信息传给地图快照，让快照不断地更新，代码如下：

```
using System.Collections;
using System.Collections.Generic;
using UnityEngine;

public class GameController : MonoBehaviour
{
    public Block[] blocks;              //存储7种方块类型
    private Block curBlock;
    private Block nextBlock;
    private bool isFirst = true;        //是否第一次生成方块
    public Map map;

    //Start is called before the first frame update
    void Start()
    {
        CreateBlock();
    }

    void Update()
    {
        MoveBlock();
    }

    //生成方块
    void CreateBlock()
    {
        //如果游戏刚开始
        if (isFirst == true)
        {
            //随机实例化一个方块预制体
            int curBlockIndex = Random.Range(0, blocks.Length);
            curBlock = Instantiate(blocks[curBlockIndex]);
            isFirst = false;

            int nextBlockIndex = Random.Range(0, blocks.Length);
            nextBlock = Instantiate(blocks[nextBlockIndex]);
            Debug.Log(curBlock.name + " " + nextBlock.name);
            //在 Console 窗口中输出 curBlock 和 nextBlock 的名称
```

```csharp
        }
        else
        {
            Destroy(curBlock.gameObject);
            curBlock = nextBlock;
            nextBlock = Instantiate(blocks[Random.Range(0, blocks.Length)]);
            SetBlockToMap(curBlock.GetCurPos());
        }
    }

    //移动和旋转方块
    void MoveBlock()
    {
        if (curBlock != null)
        {
            if (Input.GetKeyDown(KeyCode.LeftArrow))
            {
                //方块向左移动一个单位
                MoveCurBlockLeft();
            }
            else if (Input.GetKeyDown(KeyCode.RightArrow))
            {
                //方块向右移动一个单位
                MoveCurBlockRight();
            }
            else if (Input.GetKeyDown(KeyCode.DownArrow))
            {
                //方块向下移动一个单位
                MoveCurBlockDown();
            }
            else if (Input.GetKeyDown(KeyCode.UpArrow))
            {
                //方块旋转
                RotateCurBlock();
            }
        }
    }

    void MoveCurBlockLeft()
    {
        //左移
        curBlock.MoveLeft();
        //判断是否碰到墙体、其他方块
        if (!CanMoveBlock(curBlock.GetCurPos()))
        {
            //如果不能向左移动,则回到原来的位置
            curBlock.MoveRight();
            return;
        }
        //刷新地图
        map.ClearMap();
        SetBlockToMap(curBlock.GetCurPos());
        map.ShowMapInfo();
```

```cpp
    }
    void MoveCurBlockRight()
    {
        curBlock.MoveRight();
        if (!CanMoveBlock(curBlock.GetCurPos()))
        {
            curBlock.MoveLeft();
            return;
        }
        map.ClearMap();
        SetBlockToMap(curBlock.GetCurPos());
        map.ShowMapInfo();
    }

    void MoveCurBlockDown()
    {
        curBlock.MoveDown();
        if (!CanMoveBlock(curBlock.GetCurPos()))
        {
            curBlock.MoveUp();
            return;
        }
        map.ClearMap();
        SetBlockToMap(curBlock.GetCurPos());
        map.ShowMapInfo();
    }

    void RotateCurBlock()
    {
        curBlock.RotateBlock();
        if (!CanMoveBlock(curBlock.GetCurPos()))
        {
            curBlock.InverseRotateBlock();
            return;
        }
        map.ClearMap();
        SetBlockToMap(curBlock.GetCurPos());
        map.ShowMapInfo();
    }

    //判断方块是否可以移动
    bool CanMoveBlock(Vector2 curPos)
    {
        return true;
    }

    void SetBlockToMap(Vector2 curPos)
    {
        for (int blockRow = 0; blockRow < 4; blockRow++)
        {
            for (int blockCol = 0; blockCol < 4; blockCol++)
            {
```

```
            if (curBlock.shape[curBlock.GetBlockRotateState(),
              blockRow * 4 + blockCol] != 0)
            {
                map.SetMapItem((int)curPos.y - blockRow,
                          blockCol + (int)curPos.x,
                        curBlock.shape[curBlock.GetBlockRotateState(),
                          blockRow * 4 + blockCol],
                        curBlock.blockColor);
            }
        }//blockCol
    }//blockRow
}
```

（3）返回 Unity 界面，将 Map 物体赋值给 GameController 脚本，如图 6-26 所示。

（4）运行游戏，可以看到效果图如图 6-27 所示。此时按键盘上的方向键←、方向键→和方向键↓，方块便会朝着对应的方向移动；按方向键↑，方块会变换形状。

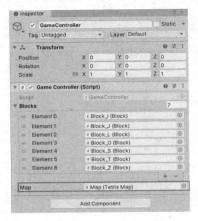
图 6-26　为 GameController 脚本赋值

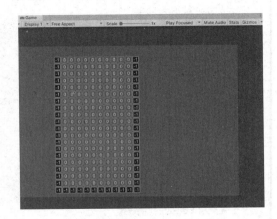
图 6-27　效果图

6.4.8　边界的判断

在移动和旋转方块时，有一个非常重要的逻辑：边界的判断。

实现思路如下：先进行移动或旋转，再判断方块是否出界。具体的判断方法是遍历方块的 shape 数组中的每个元素，如果该元素不为 0，则表示该处有一个小方块，此时需要判断是否碰到左侧墙体、右侧墙体、底部或其他方块。如果发生了碰撞，则回撤移动和旋转操作。

（1）打开脚本编辑器，先在 Map 脚本中实现一个用于获取指定位置的地图快照信息的函数 GetMapInfo()，再在 GameController 脚本中具体实现 CanMoveBlock() 函数。

Map 脚本需要添加的代码如下：

```
//获得每个方块的信息
public int GetMapInfo(int row,int col)
{
    return mapSnapshot[row, col];
}
```

GameController 脚本需要添加的代码如下：

```
//判断方块是否可以移动
bool CanMoveBlock(Vector2 curPos)
```

```csharp
{
    //遍历该方块当前形状下的数组
    for (int blockRow = 0; blockRow < 4; blockRow++)
    {
        for (int blockCol = 0; blockCol < 4; blockCol++)
        {
            if (curBlock.shape[curBlock.GetBlockRotateState(),
                blockRow * 4 + blockCol] != 0)
            {   //只判断有方块的位置
                //当碰到墙体、其他方块，或者到达底部时，方块将不能移动
                if (HitWall(blockRow, blockCol, curPos) ||
                    HitOtherBlock(blockRow, blockCol, curPos) ||
                    ArriveBottom(blockRow, blockCol, curPos))
                {
                    return false;
                }
            }
        }//blockCol
    }//blockRow
    return true;
}

//判断是否碰到其他方块
bool HitOtherBlock(int blockRow, int blockCol, Vector2 curPos)
{
    if (map.GetMapInfo((int)curPos.y - blockRow, (int)curPos.x + blockCol) == 8)
    {
        return true;
    }
    return false;
}

//判断是否碰到边界
bool HitWall(int blockRow, int blockCol, Vector2 curPos)
{
    if (map.GetMapInfo((int)curPos.y - blockRow, (int)curPos.x + blockCol) == -1)
    {
        return true;
    }
    return false;
}

//判断是否到达底部
bool ArriveBottom(int blockRow, int blockCol, Vector2 curPos)
{
    if (curPos.y - blockRow <= 0)
    {
        return true;
    }
    return false;
}
```

（2）当方块到达底部时便不能再移动了，需要把它变成固定方块，接着生成一个个新的

方块。

（3）先在 Map 脚本中实现用于设置方块固定的函数 SetFixedItem()；再在 GameController 脚本中修改 MoveCurBlockDown()函数，从而将不能再移动的方块在地图中设置为 8，并重新生成一个新的方块。

Map 脚本需要添加的代码如下：

```
public void SetFixedItem(int row, int col){
    mapSnapshot[row,col] = 8;
}
```

GameController 脚本中的代码如下：

```
using System.Collections;
using System.Collections.Generic;
using UnityEngine;

public class GameController : MonoBehaviour
{
    //……

    void MoveCurBlockDown()
    {
        curBlock.MoveDown();
        if (!CanMoveBlock(curBlock.GetCurPos()))
        {
            curBlock.MoveUp();
            SetFixedBlockAtMap(curBlock.GetCurPos());
            CreateBlock();          //重新创建一个方块
            return;
        }
        map.ClearMap();
        SetBlockToMap(curBlock.GetCurPos());
        map.ShowMapInfo();
    }

    //……

    //设置固定方块
    void SetFixedBlockAtMap(Vector2 curPos)
    {
        for (int blockRow = 0; blockRow < 4; blockRow++)
        {
            for (int blockCol = 0; blockCol < 4; blockCol++)
            {
                if (curBlock.shape[curBlock.GetBlockRotateState(),
                  blockRow * 4 + blockCol] != 0)
                {
                    map.SetFixedItem((int)curPos.y - blockRow,
                      blockCol + (int)curPos.x);
                }
            }
        }
    }
}
```

6.4.9 满行的消除

当方块到达底部后,需要再次进行判断。首先判断是否碰到了顶部,如果碰到了,则游戏结束,否则检测地图上哪一行是满行。如果有满行,则消除这一行的数据,上方固定方块数据下移。如果不满足满行的条件,则重新生成方块。

需要注意的是,当进行满行的消除、上方固定方块数据下移后,可能会产生新的满行,这是递归的过程。

(1)打开脚本编辑器,在 Map 脚本中编写用于检测满行的函数 DetectLine(),以及用于删除一行满行的函数 DelectLine()。Map 脚本需要添加的代码如下:

```csharp
//判断是否满行
public void DetectLine()
{
    int count = 0;
    for (int row = 1; row < mapRow - 1; row++)
    {
        count = 0;
        for (int col = 1; col < mapCol - 1; col++)
        {
            if (mapSnapshot[row, col] == 8)
            {   //计算每行中 8 的个数
                count++;
            }
            else
            {
                //如果该行中有一个非 8 数字,则退出该行的遍历,进行下一行的遍历
                break;
            }
        } // col
        if (count == mapCol - 2)
        {   //如果个数达到该行空格个数,则表示满行
            Debug.Log("Whole Line " + row);
            DelectLine(row);    //消除该行
        }
    }//row
}

//删除一行满行
public void DelectLine(int delectRow)
{
    for (int row = delectRow; row < mapRow - 1; row++)
    {
        for (int col = 1; col < mapCol - 1; col++)
        {
            mapSnapshot[row, col] = mapSnapshot[row + 1, col];
            backgroundObjs[row, col].GetComponent<SpriteRenderer>().color =
                backgroundObjs[row + 1, col].GetComponent<SpriteRenderer>().color;
        }
    }
    DetectLine();   //在消除该行后,需要重新进行遍历
}
```

（2）修改 GameController 脚本中的 MoveCurBlockDown()函数，使其调用在 Map 脚本中刚刚编写的函数进行满行消除判断，代码如下：

```
void MoveCurBlockDown()
{
    curBlock.MoveDown();
    if (!CanMoveBlock(curBlock.GetCurPos()))
    {
        curBlock.MoveUp();
        SetFixedBlockAtMap(curBlock.GetCurPos());
        map.DetectLine();           //进行满行判断
        CreateBlock();              //重新创建一个方块
        return;
    }
    map.ClearMap();
    SetBlockToMap(curBlock.GetCurPos());
    map.ShowMapInfo();
}
```

（3）返回 Unity 界面，单击"运行"按钮，满行消除前后的效果如图 6-28 所示。

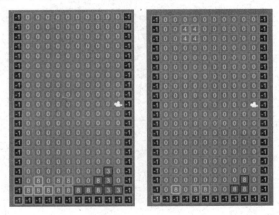

图 6-28　满行消除前后的效果

6.4.10　游戏结束的判断

遍历地图顶部的数值，如果有一个数值为 8，则表示游戏结束。

（1）在 Map 脚本中编写一个判断游戏结束的函数 DetectGameOver()。Map 脚本需要添加的代码如下：

```
public bool DetectGameOver()
{
    for (int fullCol = 1; fullCol < mapCol - 1; fullCol++)
    {
        if (mapSnapshot[mapRow - 1, fullCol] == 8)
        {
            return true;
        }
    }
    return false;
}
```

（2）打开 GameController 脚本，添加一个布尔型变量 isGameOver 来代表游戏是否结束。只有当 isGameOver 为 false 时，才可以进行方块的移动和旋转。修改 MoveCurBlockDown() 函数，在满行消除判断前先进行是否结束游戏的判断。GameController 脚本中的代码如下：

```csharp
using System.Collections;
using System.Collections.Generic;
using UnityEngine;

public class GameController : MonoBehaviour
{
    public Block[] blocks;                //存储7种方块类型
    private Block curBlock;
    private Block nextBlock;
    private bool isFirst = true;          //是否是第一次生成方块
    public Map map;
    private bool isGameOver = false;

    void Start()
    {
        CreateBlock();
    }

    void Update()
    {
        if (isGameOver == false)
        {
            MoveBlock();
        }
    }

    ……

    void MoveCurBlockDown()
    {
        curBlock.MoveDown();
        if (!CanMoveBlock(curBlock.GetCurPos()))
        {
            curBlock.MoveUp();
            SetFixedBlockAtMap(curBlock.GetCurPos());
            if (map.DetectGameOver() == true)        //游戏结束判断
            {
                Debug.Log("Game Over!");
                isGameOver = true;
                curBlock = null;
                return;
            }
            map.DetectLine();           //满行判断
            CreateBlock();              //重新生成一个方块
            return;
        }
        map.ClearMap();
        SetBlockToMap(curBlock.GetCurPos());
        map.ShowMapInfo();
```

    }
}

（3）返回 Unity 界面，单击"运行"按钮，游戏结束画面如图 6-29 所示。

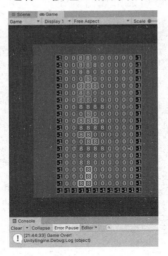

图 6-29 游戏结束画面

6.4.11 附加功能——控制方块自动下落的速度

要实现方块的自动下落，需要先实现一个节奏计时器，按照一定的时间间隔运行方块下落逻辑。

（1）在 GameController 脚本中找到监听键盘的函数 MoveBlock()，在该函数中实现如果没有按方向键↓，则每隔一段时间自动下落的功能。GameController 脚本中的代码如下：

```
using System.Collections;
using System.Collections.Generic;
using UnityEngine;

public class GameController : MonoBehaviour
{
    ……
    public float moveDownSpeed = 1f;           //控制移动的时间间隔
    private float timeCount = 0;               //用于时间计数

    ……
    //移动和旋转方块
    void MoveBlock()
    {
        if (curBlock != null)
        {
            if (Input.GetKeyDown(KeyCode.LeftArrow))
            {
                //方块向左移动一个单位
                MoveCurBlockLeft();
            }
            else if (Input.GetKeyDown(KeyCode.RightArrow))
            {
```

```csharp
            //方块向右移动一个单位
            MoveCurBlockRight();
        }
        else if (Input.GetKeyDown(KeyCode.DownArrow))
        {
            //方块向下移动一个单位
            MoveCurBlockDown();
        }
        else if (Input.GetKeyDown(KeyCode.UpArrow))
        {
            //方块旋转
            RotateCurBlock();
        }
    }

    if (TimeCountDown() == true)
    {
        MoveCurBlockDown();
    }
}

......

//计时器
bool TimeCountDown()
{
    timeCount += Time.deltaTime;
    if (timeCount > moveDownSpeed)
    {
        timeCount = 0;
        return true;
    }
    return false;
}
```

（2）返回 Unity 界面，单击"运行"按钮，此时如果不按键盘上的方向键，方块便会每隔 1 秒自动下落一个单位，而按方向键↓则可以加速它的下落。

6.4.12 附加功能——显示下一个方块的类型

需要在场景的右侧显示下一个要落下的方块的类型，这就需要使用一个 4×4 的二维数组来显示方块提示。

（1）在 GameController 脚本中编写 ShowNextBlock()函数，并在每次生成方块时调用该函数。GameController 脚本中的代码如下：

```csharp
using System.Collections;
using System.Collections.Generic;
using UnityEngine;

public class GameController : MonoBehaviour
{
```

……

```csharp
public GameObject[,] nextShowBlocks;      //下一个要显示的方块
public GameObject nextShowBlockFrame;
```

……

```csharp
//生成方块
void CreateBlock()
{
    //如果游戏刚开始
    if (isFirst == true)
    {
        //随机实例化一个方块预制体
        int curBlockIndex = Random.Range(0, blocks.Length);
        curBlock = Instantiate(blocks[curBlockIndex]);
        isFirst = false;

        int nextBlockIndex = Random.Range(0, blocks.Length);
        nextBlock = Instantiate(blocks[nextBlockIndex]);
        ShowNextBlock();            //显示下一个方块的类型
        SetBlockToMap(curBlock.GetCurPos());
    }
    else
    {
        Destroy(curBlock.gameObject);
        curBlock = nextBlock;
        nextBlock = Instantiate(blocks[Random.Range(0, blocks.Length)]);
        ShowNextBlock();            //显示下一个方块的类型
        SetBlockToMap(curBlock.GetCurPos());
    }
}
```

……

```csharp
//显示下一个方块的类型
void ShowNextBlock()
{
    if (nextShowBlocks == null)
    { //初始化显示下一个方块的图案
        nextShowBlocks = new GameObject[4, 4];
    }

    for (int row = 0; row < 4; row++)
    {
        for (int col = 0; col < 4; col++)
        {
            if (nextShowBlocks[row, col] == null)
            {
                nextShowBlocks[row, col] = Instantiate(nextShowBlockFrame,
                                    new Vector3(col + 15, -row + 10, 0),
                                    Quaternion.identity);
            }
```

```
                nextShowBlocks[row, col].GetComponent<SpriteRenderer>().color =
                    new Color(0, 0, 0, 0);  //隐藏空的方格
                if (nextBlock.shape[nextBlock.GetBlockRotateState(),
row * 4 + col] != 0)
                {  //把方格对应位置的颜色显示出来
                    nextShowBlocks[row, col].GetComponent<SpriteRenderer>()
                        .color = nextBlock.blockColor;
                }
            }//col
        }//row
    }
}
```

（2）返回 Unity 界面，将 Prefabs 文件夹中的 BlockFrame 预制体赋给 GameController 脚本中的 Next Show Block Frame，如图 6-30 所示。

（3）返回 Unity 界面，单击"运行"按钮，可以看到此时地图右侧已经显示出下一个方块的类型。图 6-31 所示为下一个方块的类型。

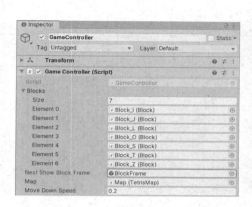 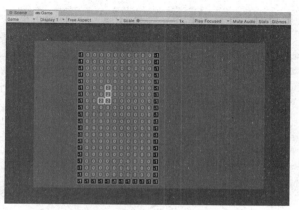

图 6-30　将 BlockFrame 预制体赋给 Next Show Block Frame　　图 6-31　下一个方块的类型

6.4.13　附加功能——优化重复代码

为了使代码看上去更加简洁、优雅，可以进行简单的重构工作。可以发现，当左移、右移、下移时，很多代码都是一模一样的，因此可以通过重构优化重复代码。

（1）找出左移、右移、下移和旋转函数中的重复代码，将它们放在单独的一个函数中，并调用这个函数来替换原来的重复代码。GameController 脚本中的代码如下：

```
using System.Collections;
using System.Collections.Generic;
using UnityEngine;

public class GameController : MonoBehaviour
{
    ……

    void MoveCurBlockLeft()
    {
        //左移
        curBlock.MoveLeft();
```

```csharp
            //判断是否碰到墙体、其他方块
            if (!CanMoveBlock(curBlock.GetCurPos()))
            {
                //如果不能向左移动，则回到原来的位置
                curBlock.MoveRight();
                return;
            }
            //刷新地图
            UpdateMap(curBlock.GetCurPos());
            //map.ClearMap();
            //SetBlockToMap(curBlock.GetCurPos());
            //map.ShowMapInfo();
        }

        void MoveCurBlockRight()
        {
            curBlock.MoveRight();
            if (!CanMoveBlock(curBlock.GetCurPos()))
            {
                curBlock.MoveLeft();
                return;
            }
            UpdateMap(curBlock.GetCurPos());
            //map.ClearMap();
            //SetBlockToMap(curBlock.GetCurPos());
            //map.ShowMapInfo();
        }

        void MoveCurBlockDown()
        {
            curBlock.MoveDown();
            if (!CanMoveBlock(curBlock.GetCurPos()))
            {
                curBlock.MoveUp();
                SetFixedBlockAtMap(curBlock.GetCurPos());
                if (map.DetectGameOver() == true)           //游戏结束的判断
                {
                    Debug.Log("Game Over!");
                    isGameOver = true;
                    curBlock = null;
                    return;
                }
                map.DetectLine();    //满行的判断
                CreateBlock();       //重新生成一个方块
                return;
            }
            UpdateMap(curBlock.GetCurPos());
            //map.ClearMap();
            //SetBlockToMap(curBlock.GetCurPos());
            //map.ShowMapInfo();
        }

        void RotateCurBlock()
```

```csharp
        {
            curBlock.RotateBlock();
            if (!CanMoveBlock(curBlock.GetCurPos()))
            {
                curBlock.InverseRotateBlock();
                return;
            }
            UpdateMap(curBlock.GetCurPos());
            //map.ClearMap();
            //SetBlockToMap(curBlock.GetCurPos());
            //map.ShowMapInfo();
        }
        ……
        void UpdateMap(Vector2 curBlockPos)
        {
            map.ClearMap();
            SetBlockToMap(curBlockPos);
            map.ShowMapInfo();
        }
```

（2）返回 Unity 界面，检查重构后是否产生了 Bug，若没有，则代码的重构完成。

6.5　本章小结

本章主要介绍了《俄罗斯方块》游戏的设计与制作，其中重点在于方块数据结构的设计、方块的旋转与平移、方块的碰撞判断，以及消除满行的实现。

6.6　练习题

1. 若以四连通域为标准（任意两个正方形小方块之间只能通过邻边实现可达），则由 4 个小方块组成的俄罗斯大方块最多有 7 种类型（与位置和朝向无关）。由 5 个小方块组成的俄罗斯大方块有多少种类型（与位置和朝向无关）？请一一画出。

2. 在进行《俄罗斯方块》游戏的平移和旋转可行性判断时采用的策略为"先斩后奏"（先操作后判断，若不满足条件再撤销），而在《推箱子》游戏中进行此类判断时则采取了相反的谨慎态度（先判断可行性，再决定是否移动），请分析二者的优点和缺点。

3. （选做）在消除满行的过程中，可以考虑另一种实现方式：只处理当前方块所在的若干行，先记录其中所有满行的行号，再同时删除，并将上方诸行同时向下移动。请问该思路是否可行，若可行，请对比分析该思路与原实现方式的优点和缺点，以及实现中需要注意的问题。提示：地图的数据结构是否需要修改，如果需要修改，应该修改成什么？为什么？

第 7 章

华容道

7.1 游戏简介

《华容道》是中国民间益智游戏，并且以变化多端、百玩不厌的特点与魔方、独立钻石棋一起被国外智力专家并称为"智力游戏界的 3 个不可思议"。它与七巧板、九连环等中国传统益智玩具的另一个代名词为"中国的难题"。《华容道》游戏的画面如图 7-1 所示。

图 7-1 《华容道》游戏的画面

《华容道》游戏取自长篇小说《三国演义》中的故事，曹操在赤壁大战中被刘备和孙权的"苦肉计"和"铁索连舟"打败，被迫退到华容道，但又遇上诸葛亮的伏兵，关羽为了报答曹操对他的恩情，明逼实让，最终帮助曹操逃出了华容道。《华容道》游戏就是依照"曹瞒兵败走华容，正与关公狭路逢。只为当初恩义重，放开金锁走蛟龙"的故事情节设计的。《华容道》游戏现在的样式是 1932 年 John Harold Fleming 在英国申请的专利，并且附有"横刀立马"的解法。

《华容道》游戏是中国人发明的，最终解法是美国人用计算机求出的。但是《华容道》游戏的数学原理到现在仍然是一个未解之谜。

7.2 游戏规则

（1）《华容道》游戏的棋盘有 20 个方块，棋盘下方有一个宽度为 2 个方格的出口，仅供曹操逃走。

（2）《华容道》游戏中的方块有 4 种：大正方形（2×2）、横长方形（2×1）、竖长方形（1×2）、小正方形（1×1）。

（3）《华容道》游戏有不同的开局，根据 5 个长方形块的放法分类，除了 5 个都竖放是不可能的，还有一横式、二横式、三横式、四横式和五横式。该游戏有"横刀立马"、"近在咫尺"、"过五关"、"水泄不通"和"小燕还巢"等玩法。图 7-2 所示为《华容道》游戏布局方法举例。

（4）棋盘上仅有 2 个空格可供棋子移动。

（5）玩家需要移动棋子，让曹操从棋盘最下方中部的 2 个空格逃出，此时游戏胜利。

图 7-2 《华容道》游戏布局方法举例

7.3 游戏开发核心思路

7.3.1 棋子类（Chess）的编写

（1）使用枚举类型编写棋子的类型。
（2）初始化、生成棋子对象。
（3）根据棋子类型为棋子的宽度和高度赋值。
（4）（后续）需要设置曹操目标点。
（5）（后续）需要编写曹操是否到达目标点的判断函数。

7.3.2 棋盘盘面布局的生成

（1）将所有棋子对象保存到数组中，并用棋子类实例化棋子对象。设定棋子的锚点在左上角。棋子的位置通过硬编码方式进行设置。
（2）用二维数组初始化棋盘快照。
棋盘快照的作用：记录当前盘面的状态，判断被选择的棋子是否可以被移动到下一个位置，以及是否已胜利等。
棋盘状态有以下几种。
- 棋盘已被棋子占用，设置 ID 为 1。
- 棋子在棋盘的边界，设置 ID 为 1。
- 棋盘没有被棋子占用，设置 ID 为 0。

注意：需要增加两行两列表示棋盘边界。

7.3.3 移动棋子的实现

1）判断选中的棋子

获取单击时鼠标指针的位置坐标（这里获得的是鼠标指针的屏幕坐标），需要将其转化为世界坐标，使鼠标指针的位置坐标与棋子的位置坐标在同一坐标系下；循环遍历所有棋子，判断单击的世界位置坐标是在哪一个棋子的坐标范围内。

设鼠标指针的位置为(M_x, M_y)，棋子的宽度为w，棋子的高度h，棋子左上角坐标为(L_x, L_y)。所以，判断鼠标指针的位置是否在棋子中的关系为$L_x<M_x<L_x+w$且$L_y-h<M_y<L_y$。满足以上条件的棋子被选中。判断选中的棋子的示意图如图7-3所示。

2）判断棋子的移动方向

通过鼠标移动棋子。判断鼠标指针移动方向的步骤如下：当按下鼠标按键时记录鼠标指针的位置P_1，当松开鼠标按键时记录鼠标指针的位置P_2，根据计算P_1与P_2的偏移量判断鼠标指针移动的方向。

图7-4所示为棋子移动方向的判断，假设单击的位置坐标为$P_1(x_1, y_1)$，松开鼠标按键时的位置坐标为$P_2(x_2, y_2)$，计算P_1与P_2的偏移量P_2-P_1。

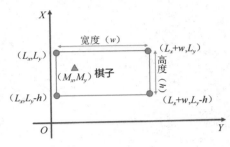

图7-3 判断选中的棋子的示意图

- 当x_2-x_1为负且$|y_2-y_1|<0.5$时，移动方向为向左。
- 当x_2-x_1为正且$|y_2-y_1|<0.5$时，移动方向为向右。
- 当y_2-y_1为负且$|x_2-x_1|<0.5$时，移动方向为向下。
- 当y_2-y_1为正且$|x_2-x_1|<0.5$时，移动方向为向上。

总之，Δx用于判断向左还是向右移动，Δy用于判断向上还是向下移动，另一条轴向的偏移量绝对值应小于0.5。

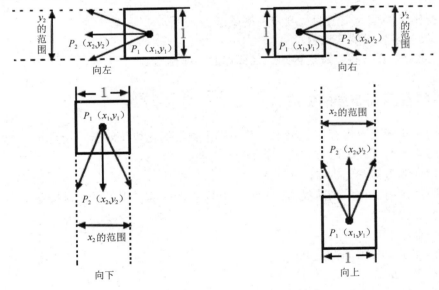

图7-4 棋子移动方向的判断

3）判断棋子的类型

根据宽度和高度不同可以将棋子分为4种类型。

- 棋子形状为正方形 1×1。
- 棋子形状为长方形 1×2。
- 棋子形状为长方形 2×1。

- 棋子形状为正方形 2×2。

4）判断棋子是否可以移动

为了判断棋子是否可以移动，需要知道棋子的移动方向、棋子的类型、棋子的位置（x,y），以及棋子在地图快照中的索引（row,col）。

下面根据棋子的类型判断棋子下一个位置在地图快照中的值。

首先，获得当前棋子在地图快照中的索引：被选中棋子在地图快照中的行索引 = 地图快照总行数 − 被选中棋子的 y 坐标−1，被选中棋子在地图快照中的列索引 = 被选中棋子的 x 坐标。

然后，获得棋子下一个位置在地图快照中的索引：若下一个位置的索引所在格子状态 = 0，则可以移动；若下一个位置的索引所在格子状态 = 1，则不可以移动。

最后，移动棋子：改变棋子在场景中的位置，刷新地图快照。

下面根据棋子的类型分析棋子的移动情况。

（1）1×1 棋子（兵）。

首先获取棋子的左上角坐标，得到棋子所在位置（x,y），然后判断移动的方向。

- 如果向右移动，则判断（$x+1,y$）位置是否为空。若为空，则将棋子的位置的 x 坐标加 1，更新棋盘的状态；若不为空，则棋子不能向右移动，如图 7-5 所示。
- 如果向左移动，则判断（$x-1,y$）位置是否为空。若为空，则将棋子所在位置的 x 坐标减 1，更新棋盘的状态；若不为空，则棋子不能向左移动，如图 7-6 所示。
- 如果向下移动，则判断（$x,y-1$）位置是否为空。若为空，则将棋子所在位置的 y 坐标减 1，更新棋盘的状态；若不为空，则棋子不能向下移动，如图 7-7 所示。
- 如果向上移动，则判断（$x,y+1$）位置是否为空。若为空，则将棋子所在位置的 y 坐标加 1，更新棋盘的状态；若不为空，则棋子不能向上移动，如图 7-8 所示。

图 7-5 兵向右移动

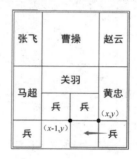
图 7-6 兵向左移动

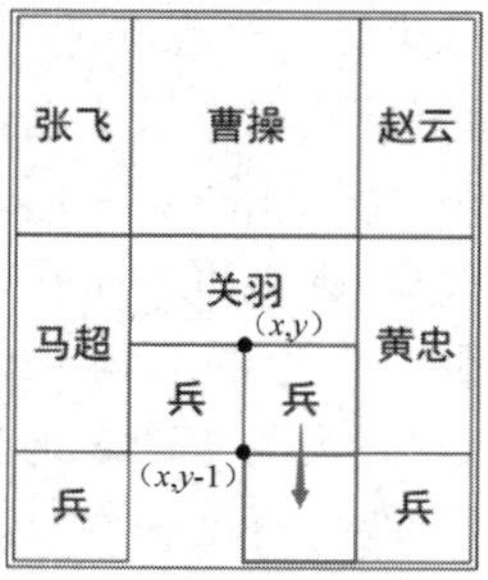

图 7-7 兵向下移动

图 7-8 兵向上移动

（2）1×2 棋子（竖向将军）。

首先获取棋子的左上角坐标，得到棋子所在位置（x,y），然后判断移动的方向。

- 如果向右移动，则判断（x+1,y）位置和（x+1,y-1）位置是否为空。若为空，则将棋子所在位置的 x 坐标加 1，更新棋盘的状态；若都不为空或只有一个为空，则棋子不能向右移动，如图 7-9 所示。
- 如果向左移动，则判断（x-1,y）位置和（x-1,y-1）位置是否为空。若为空，则将棋子所在位置的 x 坐标减 1，更新棋盘的状态；若都不为空或只有一个为空，则棋子不能向左移动，如图 7-10 所示。
- 如果向上移动，则判断（x,y+1）位置是否为空。若为空，则将棋子所在位置的 y 坐标加 1，更新棋盘的状态；若不为空，则棋子不能向上移动，如图 7-11 所示。
- 如果向下移动，则判断（x,y-1）位置是否为空。若为空，则将棋子所在位置的 y 坐标减 1，更新棋盘的状态，若不为空，则棋子不能向下移动，如图 7-12 所示。

图 7-9　竖向将军向右移动

图 7-10　竖向将军向左移动

图 7-11　竖向将军向上移动

图 7-12　竖向将军向下移动

（3）2×1 棋子（横向将军）。

首先获取棋子的左上角坐标，得到棋子所在位置（x,y），然后判断移动的方向。

- 如果向右移动，则判断（x+2,y）位置是否为空。若为空，则将棋子所在位置的 x 坐标加 1，更新棋盘的状态；若不为空，则棋子不能向右移动，如图 7-13 所示。
- 如果向左移动，则判断（x-1,y）位置是否为空。若为空，则将棋子所在位置的 x 坐标减 1，更新棋盘的状态；若不为空，则棋子不能向左移动，如图 7-14 所示。
- 如果向上移动，则判断（x,y-1）位置和（x+1,y-1）位置是否为空。若为空，则将棋子所在位置的 y 坐标加 1，更新棋盘的状态。若都不为空或只有一个为空，则棋子不能向上移动，如图 7-15 所示。
- 如果向下移动，则判断（x,y-1）位置和（x+1,y-1）位置是否为空。若为空，则将棋子所在位置的 y 坐标减 1，更新棋盘的状态；若都不为空或只有 1 个为空，则棋子不能向下移动，如图 7-16 所示。

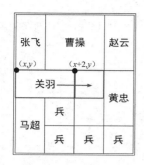

图 7-13　横向将军向右移动

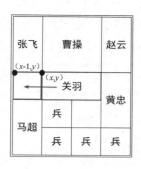

图 7-14　横向将军向左移动

图 7-15　横向将军向上移动

图 7-16　横向将军向上移动

（4）2×2 棋子（曹操）。

曹操是一个正方形的棋子，占 4 个格子。该棋子的移动与将军的移动相似。首先获取棋子的左上角坐标，得到棋子所在位置（x,y），然后判断移动的方向。

- 如果向右移动，则判断（$x+2,y$）位置和（$x+2,y-1$）位置是否为空。若为空，则将棋子所在位置的 x 坐标加 1，更新棋盘的状态；若都不为空或只有一个为空，则棋子不能向右移动，如图 7-17 所示。
- 如果向左移动，则判断（$x-1,y$）位置和（$x-1,y-1$）位置是否为空。若为空，则将棋子所在位置的 x 坐标减 1，更新棋盘的状态；若都不为空或只有一个为空，则棋子不能向右移动，如图 7-18 所示。

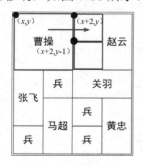

图 7-17　曹操向右移动

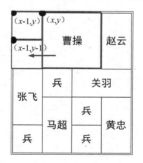

图 7-18　曹操向左移动

- 如果向上移动，则判断（$x,y+1$）位置和（$x+1,y+1$）位置是否为空。若为空，则将棋子所在位置的 y 坐标加 1，更新棋盘的状态；若都不为空或只有一个为空，则棋子不能向上移动，如图 7-19 所示。
- 如果向下移动，则判断（$x,y-2$）位置和（$x+1,y-2$）位置是否为空。若为空，则将棋子

所在位置的 y 坐标减 1，更新棋盘的状态；若都不为空或只有一个为空，则棋子不能向下移动，如图 7-20 所示。

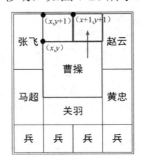
图 7-19　曹操向上移动

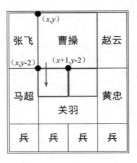
图 7-20　曹操向下移动

5）移动棋子并更新棋盘的状态

（1）移动棋子：将选中棋子的位置坐标加上移动方向向量。

（2）更新棋盘的状态：根据棋子的类型和移动方向修改有变动部分棋盘布局的状态。

7.3.4　游戏胜利的判断

当曹操移到正确的位置时，游戏结束。在最初设计游戏时，先计算好曹操最终位置的坐标，在移动过程中，判断曹操的左上角坐标是否与设置的最终位置相等，若相等，则游戏结束。

7.3.5　游戏流程图

游戏流程图如图 7-21 所示。

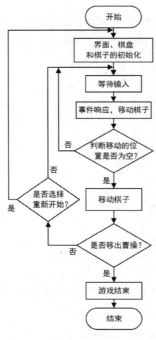

图 7-21　游戏流程图

7.4 游戏实现

7.4.1 资源的导入

（1）新建工程。首先新建名称为 HuaRongDao 的工程，在模板中选择 2D 选项，如图 7-22 所示，然后创建项目。打开工程后，按照惯例，在 Assets 文件夹下新建 Sprites 文件夹、Scripts 文件夹和 Prefabs 文件夹，这 3 个文件夹分别用来存储图片、脚本和预制体。

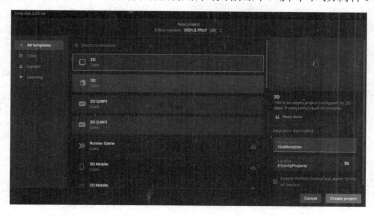

图 7-22　新建工程

（2）导入图片素材。将下载的图片素材保存到 Sprites 文件夹中，如图 7-23 所示。

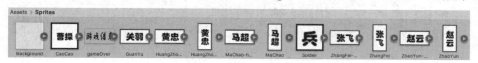

图 7-23　导入图片素材

（3）设置背景。将 Background 图片拖到 Scene 窗口中，设置该图片的大小为（2.5,2.5），位置为（3,3）。将 Order in Layer 参数设置为-10，让 Background 图片始终在最下方显示。图 7-24 所示为设置背景参数。

（4）设置摄像机。将摄像机的位置设置为（3,3），对准 Background 图片。摄像机参数如图 7-25 所示。设置好的 Game 窗口如图 7-26 所示。

图 7-24　设置背景参数

图 7-25　摄像机参数

（5）设置图片。选中 CaoCao 图片，在 Inspector 窗口中将 Pixels Per Unit 设置为 64。为了后面脚本中计算方便，还需要将 Pivot 设置为 Top Left，如图 7-27 所示，这样可以将图片的中心点移到左上角。单击 Apply 按钮保存更改。对 Sprites 文件夹下除 Background 图片和 gameOver 图片外的所有图片执行上述同样的操作。

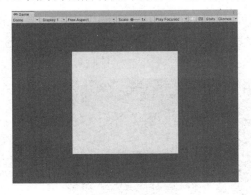
图 7-26　设置好的 Game 窗口

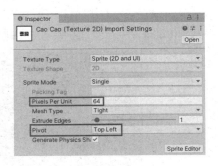
图 7-27　设置图片

（6）制作棋子预制体。先将 CaoCao 图片拖到 Scene 窗口中，再把 Hierarchy 窗口中的 CaoCao 图片拖到 Prefabs 文件夹下，使其成为一个预制体，最后把 Hierarchy 窗口中的 CaoCao 图片删除。同样，对 Sprites 文件夹下除 Background 图片和 gameOver 图片外的所有图片都执行同样的操作，如图 7-28 所示。

图 7-28　制作棋子预制体

7.4.2　棋子类的编写

（1）在 Scripts 文件夹下新建一个 C#脚本，将其重命名为 Chess 并打开。
（2）编写 Chess 脚本，用来存储棋子的属性。Chess 脚本中的代码如下：

```csharp
using System.Collections;
using System.Collections.Generic;
using UnityEngine;

//分别表示1*1、1*2、2*1、2*2 类型的棋子
public enum ChessType { Rect11, Rect12, Rect21, Rect22 }

public class Chess : MonoBehaviour
{
    public ChessType type;              //棋子的类型
    private Vector2 WH;                 //棋子的宽度和高度

    //初始化棋子
    public Chess InitChess(GameObject chessObj, Vector2 chessPos)
    {
        Chess tempChess;                //用来暂时保存生成的棋子对象
```

```
            tempChess = Instantiate(chessObj, chessPos,
                        Quaternion.identity).GetComponent<Chess>();
            tempChess.WH = SetChessWH();  //设置棋子的宽度和高度
            return tempChess;
    }

    //设置棋子的宽度和高度
    public Vector2 SetChessWH()
    {
        //宽度
        if (type == ChessType.Rect11 || type == ChessType.Rect12)
        {
            WH.x = 1;
        }
        else
        {
            WH.x = 2;
        }

        //高度
        if (type == ChessType.Rect11 || type == ChessType.Rect21)
        {
            WH.y = 1;
        }
        else
        {
            WH.y = 2;
        }
        return WH;
    }
}
```

（3）返回 Unity 界面，选中 Prefabs 文件夹下的所有预制体，在 Inspector 窗口中单击下方的 Add Component 按钮，搜索到 Chess，这样便可以为所有的棋子预制体都加上 Chess 脚本。

（4）根据棋子的类型修改 Type 属性。修改每个棋子预制体的 Chess 脚本的 Type 属性，如图 7-29 所示，如 CaoCao 为 Rect22，GuanYu 为 Rect21，以此类推。

图 7-29　修改 Type 属性

7.4.3　棋盘布局的生成

（1）在编写好棋子类后，只需要实例化棋子就可以在游戏中生成棋子。接下来在 Scripts 文件夹下创建一个名称为 GameController 的脚本用来使用棋盘式布局实例化棋子，并将棋子放到 Scene 窗口中。GameController 脚本中的代码如下：

```
using System.Collections;
using System.Collections.Generic;
using UnityEngine;
```

```csharp
public class GameController : MonoBehaviour
{
    public GameObject[] soliders = new GameObject[4];
    public GameObject huangZhong;
    public GameObject maChao;
    public GameObject zhangFei;
    public GameObject zhaoYun;
    public GameObject guanYu;
    public GameObject caoCao;
    public GameObject huangZhong_horizontal;
    public GameObject maChao_horizontal;
    public GameObject zhangFei_horizontal;
    public GameObject zhaoYun_horizontal;

    private Chess[] chessObjs;                //保存棋子实例化后的对象

    void Start()
    {
        InitGame();                           //初始化棋盘盘面
    }

    //初始化棋盘盘面
    void InitGame()
    {
        chessObjs = new Chess[10];
        chessObjs[0] = soliders[0].GetComponent<Chess>()
                    .InitChess(soliders[0], new Vector2(1, 1));    //兵
        chessObjs[1] = soliders[1].GetComponent<Chess>()
                    .InitChess(soliders[1], new Vector2(4, 1));
        chessObjs[2] = soliders[2].GetComponent<Chess>()
                    .InitChess(soliders[2], new Vector2(2, 2));
        chessObjs[3] = soliders[3].GetComponent<Chess>()
                    .InitChess(soliders[3], new Vector2(3, 2));
        chessObjs[4] = huangZhong.GetComponent<Chess>()
                    .InitChess(huangZhong, new Vector2(1, 3));     //黄忠
        chessObjs[5] = maChao.GetComponent<Chess>()
                    .InitChess(maChao, new Vector2(1, 5));         //马超
        chessObjs[6] = zhangFei.GetComponent<Chess>()
                    .InitChess(zhangFei, new Vector2(4, 3));       //张飞
        chessObjs[7] = zhaoYun.GetComponent<Chess>()
                    .InitChess(zhaoYun, new Vector2(4, 5));        //赵云
        chessObjs[8] = guanYu.GetComponent<Chess>()
                    .InitChess(guanYu, new Vector2(2, 3));         //关羽
        chessObjs[9] = caoCao.GetComponent<Chess>()
                    .InitChess(caoCao, new Vector2(2, 5));         //曹操
    }
}
```

（2）为 GameController 脚本赋值。返回 Unity 界面，在 Hierarchy 窗口中新建一个空物体并将其命名为 GameController，把 GameController 脚本附加给该物体。把 Prefabs 文件夹下的所有预制体一一对应地赋给 GameController 脚本中的 GameObject，如图 7-30 所示。

（3）单击"运行"按钮，可以看到棋盘盘面已经生成，如图 7-31 所示。

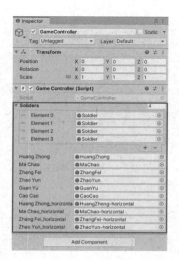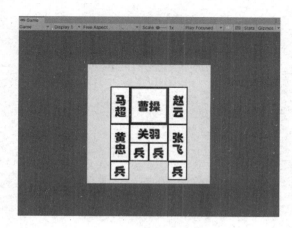

图 7-30　为 GameController 脚本中的 GameObject 赋值　　图 7-31　棋盘盘面

（4）在生成棋盘盘面后，实现棋盘快照的功能。打开 GameController 脚本继续编写代码。GameController 脚本中的代码如下：

```
using System.Collections;
using System.Collections.Generic;
using UnityEngine;
using TMPro;

public class GameController : MonoBehaviour
{
    ……

    private int[,] mapState;                          //棋盘盘面的快照数组
    private int row, col;                             //快照的行、列的数量

    public TextMeshProUGUI stateInfoText;             //快照 Text
    public TextMeshProUGUI[,] stateTextes;            //快照 Text 数组
    public Canvas canvas;                             //绘制 UI

    void Start()
    {
        InitGame();                                   //初始化棋盘盘面
        InitMapState();                               //初始化棋盘快照
        ShowChessBoardState();                        //显示棋盘快照
    }

    ……

    //初始化棋盘快照
    void InitMapState()
    {
        mapState = new int[7, 6]{
            {1,1,1,1,1,1},
            {1,1,1,1,1,1},
            {1,1,1,1,1,1},
            {1,1,1,1,1,1},
```

```
            {1,1,1,1,1,1},
            {1,1,0,0,1,1},
            {1,1,1,1,1,1}
        };
        row = mapState.GetLength(0);                    //获得行数
        col = mapState.GetLength(1);                    //获得列数
    }

    //显示棋盘快照
    void ShowChessBoardState()
    {
        if (stateTextes == null)
        {
            stateTextes = new TextMeshProUGUI[row, col];    //分配内存空间
        }
        for (int rowIndex = 0; rowIndex < row; rowIndex++)
        {
            for (int colIndex = 0; colIndex < col; colIndex++)
            {
                if (stateTextes[rowIndex, colIndex] == null)
                {
                    //实例化一个 Text 对象
                    stateTextes[rowIndex, colIndex] = Instantiate(stateInfoText);
                    //挂载在 Canvas 下
                    stateTextes[rowIndex, colIndex].transform
                        .SetParent(canvas.transform);
                    stateTextes[rowIndex, colIndex]
                        .GetComponent<RectTransform>()
                        .position = new Vector3(colIndex * 15,
                                        -rowIndex * 15, 0);
                }
                stateTextes[rowIndex, colIndex].text = mapState[rowIndex, colIndex].ToString();
            }//colIndex
        }//rowIndex
    }
```

图 7-32　设置 Text 参数

（5）设置 Text 参数。返回 Unity 界面，在主菜单栏中选择 GameObject→UI 命令，选择 Text 选项栏中的 TextMeshPro 选项，新建一个 Text，可以看到在 Hierarchy 窗口中自动生成了一个 Canvas 和一个 Text(TMP)。将 Text Input 中的文字删除，颜色改成红色，如图 7-32 所示。

（6）为 GameController 脚本赋值。将该 Text(TMP) 赋给 GameController 脚本中的 State Info Text 变量，将 Canvas 赋给 GameController 脚本的 Canvas 变量，如图 7-33 所示。

（7）单击"运行"按钮，可以看到打印的棋盘快照，如图 7-34 所示。

图 7-33 为 GameController 脚本赋值

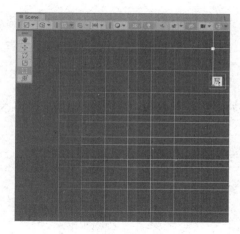

图 7-34 打印的棋盘快照

7.4.4 移动棋子——选择棋子

1. 选择棋子

（1）在 Chess 脚本中添加一个返回棋子的宽度和高度的函数 GetChessWH()。Chess 脚本需要添加的代码如下：

```
public Vector2 GetChessWH(){
    return WH;
}
```

（2）在 GameController 脚本中添加返回棋子数组和特定棋子的函数。GameController 脚本需要添加的代码如下：

```
//返回棋子数组
public Chess[] GetChessObjs(){
return chessObjs;
}

//返回特定棋子
public Chess GetChessObj(int i){
return chessObjs[i];
}
```

（3）在 Scripts 文件夹下新建一个脚本，将该脚本命名为 MouseEvent，并实现选择棋子的功能。MouseEvent 脚本中的代码如下：

```
using System.Collections;
using System.Collections.Generic;
using UnityEngine;

public class MouseEvent : MonoBehaviour
{
    Vector2 mouseBtnDownPos;           //保存按下鼠标按键时在世界坐标系中的位置
    Chess selectedChess;               //保存被选中的棋子
    public GameController gameController;
```

```
    void Update()
    {
        SelectingChess();
    }

    void SelectingChess()
    {
        if (Input.GetMouseButtonDown(0))
        {     //按下鼠标左键时
            mouseBtnDownPos = Camera.main
                            .ScreenToWorldPoint(Input.mousePosition);
            Chess selectingChess;                    //暂时保存棋子
            Vector2 chessLeftAnchor;                 //表示被遍历到的棋子的左上角坐标
            Vector2 chessWH;                         //当前被遍历到的棋子的宽度和高度

            for (int i = 0; i < gameController.GetChessObjs().Length; i++)
            {
                selectingChess = gameController.GetChessObj(i);
                chessLeftAnchor = selectingChess.transform.position;
                chessWH = selectingChess.GetChessWH();
                if ((mouseBtnDownPos.x > chessLeftAnchor.x &&
                    mouseBtnDownPos.x < chessLeftAnchor.x + chessWH.x) &&
                    (mouseBtnDownPos.y < chessLeftAnchor.y &&
                    mouseBtnDownPos.y > chessLeftAnchor.y - chessWH.y))
                {
                    selectedChess = selectingChess;
                    Debug.Log(selectedChess.name);
                    //打印选中的棋子的名字
                    break;
                }
                else
                {
                    selectingChess = null;
                    selectedChess = null;
                }//if
            }//for
        }
    }
}
```

（4）为 MouseEvent 脚本赋值。返回 Unity 界面，为 GameController 物体附加 MouseEvent 脚本，并把 GameController 脚本赋给 MouseEvent 脚本中的 Game Controller 变量，如图 7-35 所示。

（5）单击"运行"按钮，运行游戏。单击棋盘上的棋子，Console 窗口中会打印出选中的棋子的名字，如图 7-36 所示。

2．判断拖动方向

（1）在 MouseEvent 脚本中添加一个松开鼠标按键时的位置向量，用来判断拖动方向，并用一个枚举型变量来保存这个方向，代码如下：

图 7-35　为 MouseEvent 脚本赋值

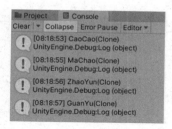

图 7-36　打印出选中的棋子的名字

```
using System.Collections;
using System.Collections.Generic;
using UnityEngine;

public enum MoveDirection { Left, Right, Up, Down }    //表示移动方向

public class MouseEvent : MonoBehaviour
{
    ……

    Vector2 mouseBtnUpPos;      //保存松开鼠标按键时在世界坐标系中的位置
    MoveDirection moveDir;      //移动方向

    void Update()
    {
        SelectingChess();
        DragChess();
    }

    ……

    //拖动棋子
    void DragChess()
    {
        //已经选择了一个棋子时
        if (selectedChess != null && Input.GetMouseButtonUp(0))
        {
            //把鼠标指针位置转换为世界坐标
            mouseBtnUpPos = Camera.main
                     .ScreenToWorldPoint(Input.mousePosition);
            if (mouseBtnDownPos.x > mouseBtnUpPos.x &&
               Mathf.Abs(mouseBtnUpPos.y - mouseBtnDownPos.y) < 0.5f)
            {
                moveDir = MoveDirection.Left;        //左移
            }
```

```
            if (mouseBtnDownPos.x < mouseBtnUpPos.x &&
                Mathf.Abs(mouseBtnUpPos.y - mouseBtnDownPos.y) < 0.5f)
            {
                moveDir = MoveDirection.Right;          //右移
            }
            if (mouseBtnDownPos.y > mouseBtnUpPos.y &&
                Mathf.Abs(mouseBtnUpPos.x - mouseBtnDownPos.x) < 0.5f)
            {
                moveDir = MoveDirection.Down;           //下移
            }
            if (mouseBtnDownPos.y < mouseBtnUpPos.y &&
                Mathf.Abs(mouseBtnUpPos.x - mouseBtnDownPos.x) < 0.5f)
            {
                moveDir = MoveDirection.Up;             //上移
            }
            Debug.Log(selectedChess.name + " Move " + moveDir);
        }
    }
}
```

（2）判断拖动方向。返回 Unity 界面，单击"运行"按钮，此时可以看到选中一个棋子，并且在拖动棋子时 Console 窗口中会输出拖动方向，如图 7-37 所示。

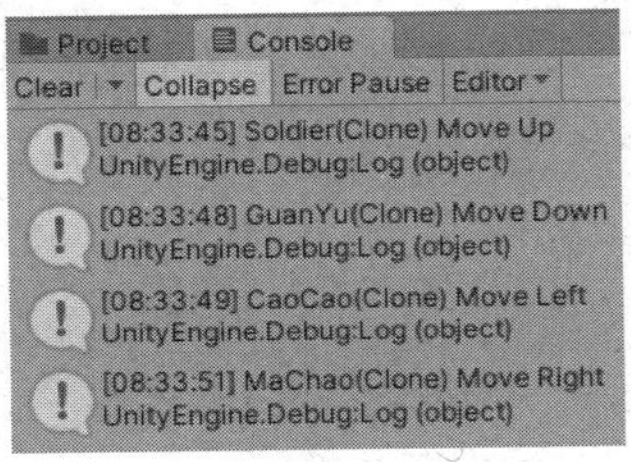

图 7-37　判断拖动方向

3．编写移动框架

（1）实现判断棋子类型的功能。在判断棋子的移动方向后，还需要判断棋子的类型，这对后面判断棋子是否可以移动十分重要，可以用 switch 语句来实现这项功能。在 MouseEvent 脚本中编写判断棋子类型及其移动的框架。MouseEvent 脚本中的代码如下：

```
using System.Collections;
using System.Collections.Generic;
using UnityEngine;

public enum MoveDirection { Left, Right, Up, Down }     //表示移动方向

public class MouseEvent : MonoBehaviour
{
    ……

    //拖动棋子
    void DragChess()
    {
        //已经选择了一个棋子时
        if (selectedChess != null && Input.GetMouseButtonUp(0))
        {
            ……
            //Debug.Log(selectedChess.name + " Move " + moveDir);
            Move();
        }
    }
```

```csharp
//移动框架
void Move()
{
    switch (moveDir)
    {
        case MoveDirection.Left:
            MoveLeft();         //左移
            break;
        case MoveDirection.Right:
            MoveRight();        //右移
            break;
        case MoveDirection.Up:
            MoveUp();           //上移
            break;
        case MoveDirection.Down:
            MoveDown();         //下移
            break;
    }
}

//左移
void MoveLeft()
{
    switch (selectedChess.type)
    {
        case ChessType.Rect11:
            Debug.Log("Rect11 MoveLeft");
            break;
        case ChessType.Rect12:
            Debug.Log("Rect12 MoveLeft");
            break;
        case ChessType.Rect21:
            Debug.Log("Rect21 MoveLeft");
            break;
        case ChessType.Rect22:
            Debug.Log("Rect22 MoveLeft");
            break;
    }
}

//右移
void MoveRight()
{
    switch (selectedChess.type)
    {
        case ChessType.Rect11:
            Debug.Log("Rect11 MoveRight");
            break;
        case ChessType.Rect12:
            Debug.Log("Rect12 MoveRight");
```

```csharp
            break;
        case ChessType.Rect21:
            Debug.Log("Rect21 MoveRight");
            break;
        case ChessType.Rect22:
            Debug.Log("Rect22 MoveRight");
            break;
    }
}

//上移
void MoveUp()
{
    switch (selectedChess.type)
    {
        case ChessType.Rect11:
            Debug.Log("Rect11 MoveUp");
            break;
        case ChessType.Rect12:
            Debug.Log("Rect12 MoveUp");
            break;
        case ChessType.Rect21:
            Debug.Log("Rect21 MoveUp");
            break;
        case ChessType.Rect22:
            Debug.Log("Rect22 MoveUp");
            break;
    }
}

//下移
void MoveDown()
{
    switch (selectedChess.type)
    {
        case ChessType.Rect11:
            Debug.Log("Rect11 MoveDown");
            break;
        case ChessType.Rect12:
            Debug.Log("Rect12 MoveDown");
            break;
        case ChessType.Rect21:
            Debug.Log("Rect21 MoveDown");
            break;
        case ChessType.Rect22:
            Debug.Log("Rect22 MoveDown");
            break;
    }
}
```

（2）判断棋子的类型。返回 Unity 界面，单击"运行"按钮，此时拖动棋子可以看到，Console 窗口中打印调试信息时已经完成了棋子类型的判断，如图 7-38 所示。

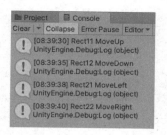

图 7-38 判断棋子的类型

7.4.5 移动棋子——移动 1×1 棋子

接下来根据棋子的类型具体实现移动棋子的功能。本节主要介绍移动 1×1 棋子。

（1）返回脚本编辑器，打开 GameController 脚本，添加几个函数用于返回所有棋盘快照，根据行、列索引返回棋盘快照信息，以及更新棋盘快照。GameController 脚本中的代码如下：

```
using System.Collections;
using System.Collections.Generic;
using UnityEngine;
using TMPro;

public class GameController : MonoBehaviour
{
    ......

    //返回棋盘快照
    public int[,] GetMapState()
    {
        return mapState;
    }

    //根据行、列索引返回棋盘快照信息
    public int GetMapStateInfo(int row, int col)
    {
        return mapState[row, col];
    }

    //更新棋盘快照
    public void UpdateMapState(int chessCurRowIndex, int chessCurColIndex,
    ChessType type, MoveDirection dir)
    {
        switch (type)
        {
            case ChessType.Rect11:
                ModifyMapStateRect11(chessCurRowIndex, chessCurColIndex, dir);
                break;
            case ChessType.Rect12:
                break;
            case ChessType.Rect21:
                break;
            case ChessType.Rect22:
                break;
```

```
        }
        ShowChessBoardState();    //刷新棋盘快照的显示
    }

    //在移动 1*1 棋子后,更新棋盘快照
    void ModifyMapStateRect11(int row, int col, MoveDirection dir)
    {
        switch (dir)
        {
            case MoveDirection.Left:
                mapState[row, col] = 0;
                mapState[row, col - 1] = 1;
                break;
            case MoveDirection.Right:
                mapState[row, col] = 0;
                mapState[row, col + 1] = 1;
                break;
            case MoveDirection.Up:
                mapState[row, col] = 0;
                mapState[row - 1, col] = 1;
                break;
            case MoveDirection.Down:
                mapState[row, col] = 0;
                mapState[row + 1, col] = 1;
                break;
        }
    }
}
```

(2) 打开 MouseEvent 脚本。添加一个结构体,用于保存棋子对应的地图快照的索引号。在移动棋子后,需要同时更新棋盘快照。MouseEvent 脚本中的代码如下:

```
using System.Collections;
using System.Collections.Generic;
using UnityEngine;

public enum MoveDirection { Left, Right, Up, Down }     //表示移动方向

//保存棋子对应的地图快照索引号
public struct ChessIndex
{
    public int rowIndex;        //行
    public int colIndex;        //列
}

public class MouseEvent : MonoBehaviour
{
    ......

    //第一个是棋子移动前的索引号,后两个是棋子移动后的索引号
    //1*1 棋子的移动只需要用到前两个
    ChessIndex chessCurIndex, chessNextIndex0, chessNextIndex1;
    private Vector2 movePos;    //表示移动时位置的改变量
```

```csharp
void Move()
{
    //根据公式获得棋子的索引号
    chessCurIndex.rowIndex = gameController.GetMapState().GetLength(0)
                             -(int)selectedChess.transform.position.y
                             -1;
    chessCurIndex.colIndex = (int)selectedChess.transform.position.x;

    switch (moveDir)
    {
        case MoveDirection.Left:
            movePos = new Vector3(-1, 0, 0);
            MoveLeft(chessCurIndex);
            break;
        case MoveDirection.Right:
            movePos = new Vector3(1, 0, 0);
            MoveRight(chessCurIndex);
            break;
        case MoveDirection.Up:
            movePos = new Vector3(0, 1, 0);
            MoveUp(chessCurIndex);
            break;
        case MoveDirection.Down:
            movePos = new Vector3(0, -1, 0);
            MoveDown(chessCurIndex);
            break;
    }
}

//左移
void MoveLeft(ChessIndex chessIndex)
{
    switch (selectedChess.type)
    {
        case ChessType.Rect11:
            //获得棋子移动后的快照索引号
            chessNextIndex0.rowIndex = chessCurIndex.rowIndex;
            chessNextIndex0.colIndex = chessCurIndex.colIndex - 1;
            //移动1*1棋子
            MoveRect11(chessCurIndex, chessNextIndex0, movePos);
            Debug.Log("Rect11 MoveLeft");
            break;
        case ChessType.Rect12:
            Debug.Log("Rect12 MoveLeft");
            break;
        case ChessType.Rect21:
            Debug.Log("Rect21 MoveLeft");
            break;
        case ChessType.Rect22:
            Debug.Log("Rect22 MoveLeft");
            break;
    }
}
```

```csharp
        }

        //右移
        void MoveRight(ChessIndex chessIndex)
        {
            switch (selectedChess.type)
            {
                case ChessType.Rect11:
                    chessNextIndex0.rowIndex = chessCurIndex.rowIndex;
                    chessNextIndex0.colIndex = chessCurIndex.colIndex + 1;
                    MoveRect11(chessCurIndex, chessNextIndex0, movePos);
                    Debug.Log("Rect11 MoveRight");
                    break;
                case ChessType.Rect12:
                    Debug.Log("Rect12 MoveRight");
                    break;
                case ChessType.Rect21:
                    Debug.Log("Rect21 MoveRight");
                    break;
                case ChessType.Rect22:
                    Debug.Log("Rect22 MoveRight");
                    break;
            }
        }

        //上移
        void MoveUp(ChessIndex chessIndex)
        {
            switch (selectedChess.type)
            {
                case ChessType.Rect11:
                    chessNextIndex0.rowIndex = chessCurIndex.rowIndex - 1;
                    chessNextIndex0.colIndex = chessCurIndex.colIndex;
                    MoveRect11(chessCurIndex, chessNextIndex0, movePos);
                    Debug.Log("Rect11 MoveUp");
                    break;
                case ChessType.Rect12:
                    Debug.Log("Rect12 MoveUp");
                    break;
                case ChessType.Rect21:
                    Debug.Log("Rect21 MoveUp");
                    break;
                case ChessType.Rect22:
                    Debug.Log("Rect22 MoveUp");
                    break;
            }
        }

        //下移
        void MoveDown(ChessIndex chessIndex)
        {
            switch (selectedChess.type)
            {
```

```
                case ChessType.Rect11:
                    chessNextIndex0.rowIndex = chessCurIndex.rowIndex + 1;
                    chessNextIndex0.colIndex = chessCurIndex.colIndex;
                    MoveRect11(chessCurIndex, chessNextIndex0, movePos);
                    Debug.Log("Rect11 MoveDown");
                    break;
                case ChessType.Rect12:
                    Debug.Log("Rect12 MoveDown");
                    break;
                case ChessType.Rect21:
                    Debug.Log("Rect21 MoveDown");
                    break;
                case ChessType.Rect22:
                    Debug.Log("Rect22 MoveDown");
                    break;
            }
        }

        //移动 1*1 棋子
        void MoveRect11(ChessIndex chessCurIndex, ChessIndex chessNextIndex0, Vector3 movePos)
        {
            if (gameController.GetMapStateInfo(chessNextIndex0.rowIndex,
                                               chessNextIndex0.colIndex) == 0)
            {
                selectedChess.transform.position += movePos;     //移动棋子
                //更新地图快照
                gameController.UpdateMapState(chessCurIndex.rowIndex,
                                              chessCurIndex.colIndex,
                                              selectedChess.type, moveDir);
            }
        }
    }
```

（3）1×1 棋子的移动效果。返回 Unity 界面，单击"运行"按钮，1×1 棋子的移动已经实现，并且每移动一次，棋盘快照都会跟着更新一次，如图 7-39 所示。

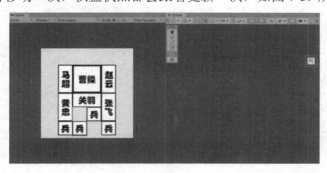

彩色图

图 7-39　1×1 棋子的移动效果

7.4.6　移动棋子——移动 1×2 棋子

相比 1×1 棋子，在移动 1×2 棋子时需要考虑棋子自身的竖向距离，更改棋盘快照时的索引号会和 1×1 棋子的情况有所不同，上面已具体分析过，此处不再赘述。

（1）在 GameController 脚本中实现 1×2 棋子移动后棋盘快照的更新。GameController 脚本中的代码如下：

```csharp
using System.Collections;
using System.Collections.Generic;
using UnityEngine;
using TMPro;

public class GameController : MonoBehaviour
{
    ……

    //更新棋盘快照
    public void UpdateMapState(int chessCurRowIndex, int chessCurColIndex,
        ChessType type, MoveDirection dir)
    {
        switch (type)
        {
            case ChessType.Rect11:
                ModifyMapStateRect11(chessCurRowIndex, chessCurColIndex, dir);
                break;
            case ChessType.Rect12:
                ModifyMapStateRect12(chessCurRowIndex, chessCurColIndex, dir);
                break;
            case ChessType.Rect21:
                break;
            case ChessType.Rect22:
                break;
        }
        ShowChessBoardState();   //刷新棋盘快照的显示
    }

    ……

    //1*2 棋子移动后，更新棋盘快照
    void ModifyMapStateRect12(int row, int col, MoveDirection dir)
    {
        switch (dir)
        {
            case MoveDirection.Left:
                mapState[row, col] = 0;
                mapState[row + 1, col] = 0;
                mapState[row, col - 1] = 1;
                mapState[row + 1, col - 1] = 1;
                break;
            case MoveDirection.Right:
                mapState[row, col] = 0;
                mapState[row + 1, col] = 0;
                mapState[row, col + 1] = 1;
                mapState[row + 1, col + 1] = 1;
                break;
            case MoveDirection.Up:
                mapState[row - 1, col] = 1;
```

```
                mapState[row + 1, col] = 0;
                break;
            case MoveDirection.Down:
                mapState[row, col] = 0;
                mapState[row + 2, col] = 1;
                break;
        }
    }
}
```

（2）在 MouseEvent 脚本中仿照 1×1 棋子移动的实现逻辑编写代码。MouseEvent 脚本中的代码如下：

```
using System.Collections;
using System.Collections.Generic;
using UnityEngine;

……

public class MouseEvent : MonoBehaviour
{
    ……

    //左移
    void MoveLeft(ChessIndex chessIndex)
    {
        switch (selectedChess.type)
        {
            ……
            case ChessType.Rect12:
                chessNextIndex0.rowIndex = chessCurIndex.rowIndex;
                chessNextIndex0.colIndex = chessCurIndex.colIndex - 1;
                chessNextIndex1.rowIndex = chessCurIndex.rowIndex + 1;
                chessNextIndex1.colIndex = chessCurIndex.colIndex - 1;
                MoveRect12(chessCurIndex, chessNextIndex0, chessNextIndex1, movePos);
                Debug.Log("Rect12 MoveLeft");
                break;
            ……
        }
    }

    //右移
    void MoveRight(ChessIndex chessIndex)
    {
        switch (selectedChess.type)
        {
            ……
            case ChessType.Rect12:
                chessNextIndex0.rowIndex = chessCurIndex.rowIndex;
                chessNextIndex0.colIndex = chessCurIndex.colIndex + 1;
                chessNextIndex1.rowIndex = chessCurIndex.rowIndex + 1;
                chessNextIndex1.colIndex = chessCurIndex.colIndex + 1;
                MoveRect12(chessCurIndex, chessNextIndex0, chessNextIndex1, movePos);
                Debug.Log("Rect12 MoveRight");
```

```csharp
            break;
        ......
    }
}

//上移
void MoveUp(ChessIndex chessIndex)
{
    switch (selectedChess.type)
    {
        ......
        case ChessType.Rect12:
            chessNextIndex0.rowIndex = chessCurIndex.rowIndex - 1;
            chessNextIndex0.colIndex = chessCurIndex.colIndex;
            MoveRect12(chessCurIndex, chessNextIndex0, chessNextIndex1, movePos);
            Debug.Log("Rect12 MoveUp");
            break;
        ......
    }
}

//下移
void MoveDown(ChessIndex chessIndex)
{
    switch (selectedChess.type)
    {
        ......
        case ChessType.Rect12:
            chessNextIndex0.rowIndex = chessCurIndex.rowIndex + 2;
            chessNextIndex0.colIndex = chessCurIndex.colIndex;
            MoveRect12(chessCurIndex, chessNextIndex0, chessNextIndex1, movePos);
            Debug.Log("Rect12 MoveDown");
            break;
        ......
    }
}

......

//移动1*2棋子
void MoveRect12(ChessIndex chessCurIndex, ChessIndex chessNextIndex0,
          ChessIndex chessNextIndex1, Vector3 movePos)
{
    //左移或右移
    if (moveDir == MoveDirection.Left || moveDir == MoveDirection.Right)
    {
        if (gameController.GetMapStateInfo(chessNextIndex0.rowIndex,
              chessNextIndex0.colIndex) == 0 &&
              gameController.GetMapStateInfo(chessNextIndex1.rowIndex,
              chessNextIndex1.colIndex) == 0)
        {
            selectedChess.transform.position += movePos;
            gameController.UpdateMapState(chessCurIndex.rowIndex,
```

```
                                        chessCurIndex.colIndex,
                                        selectedChess.type, moveDir);
            }
        }
        //上移或下移
        else if (moveDir == MoveDirection.Up || moveDir == MoveDirection.Down)
        {
            if (gameController.GetMapStateInfo(chessNextIndex0.rowIndex,
                chessNextIndex0.colIndex) == 0)
            {
                selectedChess.transform.position += movePos;
                gameController.UpdateMapState(chessCurIndex.rowIndex,
                    chessCurIndex.colIndex, selectedChess.type, moveDir);
            }
        }
    }
}
```

（3）1×2 棋子的移动效果。返回 Unity 界面，单击"运行"按钮，此时 1×2 棋子已经可以移动，并且棋盘快照也会实时更新，如图 7-40 所示。

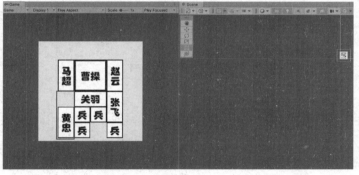

彩色图

图 7-40　1×2 棋子的移动效果

7.4.7　移动棋子——移动 2×1 棋子

在移动 2×1 棋子时需要考虑棋子自身的横向距离。

（1）在 GameController 脚本中实现 2×1 棋子移动后棋盘快照的更新。GameController 脚本中的代码如下：

```
using System.Collections;
using System.Collections.Generic;
using UnityEngine;
using TMPro;

public class GameController : MonoBehaviour
{
    ......

    //更新棋盘快照
    public void UpdateMapState(int chessCurRowIndex, int chessCurColIndex,
ChessType type, MoveDirection dir)
```

```
    {
        switch (type)
        {
            case ChessType.Rect11:
                ModifyMapStateRect11(chessCurRowIndex, chessCurColIndex, dir);
                break;
            case ChessType.Rect12:
                ModifyMapStateRect12(chessCurRowIndex, chessCurColIndex, dir);
                break;
            case ChessType.Rect21:
                ModifyMapStateRect21(chessCurRowIndex, chessCurColIndex, dir);
                break;
            case ChessType.Rect22:
                break;
        }
        ShowChessBoardState();    //刷新棋盘快照的显示
    }

    ……

    //2*1 棋子移动后，更新棋盘快照
    void ModifyMapStateRect21(int row, int col, MoveDirection dir)
    {
        switch (dir)
        {
            case MoveDirection.Left:
                mapState[row, col - 1] = 1;
                mapState[row, col + 1] = 0;
                break;
            case MoveDirection.Right:
                mapState[row, col] = 0;
                mapState[row, col + 2] = 1;
                break;
            case MoveDirection.Up:
                mapState[row, col] = 0;
                mapState[row, col + 1] = 0;
                mapState[row - 1, col] = 1;
                mapState[row - 1, col + 1] = 1;
                break;
            case MoveDirection.Down:
                mapState[row, col] = 0;
                mapState[row, col + 1] = 0;
                mapState[row + 1, col] = 1;
                mapState[row + 1, col + 1] = 1;
                break;
        }
    }
}
```

（2）仿照前面的步骤，在 MouseEvent 脚本中实现 2×1 棋子的移动。MouseEvent 脚本中的代码如下：

```
using System.Collections;
using System.Collections.Generic;
using UnityEngine;
```

```csharp
……

public class MouseEvent : MonoBehaviour
{
    ……

    //左移
    void MoveLeft(ChessIndex chessIndex)
    {
        switch (selectedChess.type)
        {
            ……
            case ChessType.Rect21:
                chessNextIndex0.rowIndex = chessCurIndex.rowIndex;
                chessNextIndex0.colIndex = chessCurIndex.colIndex - 1;
                MoveRect21(chessCurIndex, chessNextIndex0, chessNextIndex1, movePos);
                Debug.Log("Rect21 MoveLeft");
                break;
            ……
        }
    }

    //右移
    void MoveRight(ChessIndex chessIndex)
    {
        switch (selectedChess.type)
        {
            ……
            case ChessType.Rect21:
                chessNextIndex0.rowIndex = chessCurIndex.rowIndex;
                chessNextIndex0.colIndex = chessCurIndex.colIndex + 2;
                MoveRect21(chessCurIndex, chessNextIndex0, chessNextIndex1,         movePos);
                Debug.Log("Rect21 MoveRight");
                break;
            ……
        }
    }

    //上移
    void MoveUp(ChessIndex chessIndex)
    {
        switch (selectedChess.type)
        {
            ……
            case ChessType.Rect21:
                chessNextIndex0.rowIndex = chessCurIndex.rowIndex - 1;
                chessNextIndex0.colIndex = chessCurIndex.colIndex;
                chessNextIndex1.rowIndex = chessCurIndex.rowIndex - 1;
                chessNextIndex1.colIndex = chessCurIndex.colIndex + 1;
                MoveRect21(chessCurIndex, chessNextIndex0, chessNextIndex1, movePos);
                Debug.Log("Rect21 MoveUp");
                break;
```

```csharp
            ……
        }
    }

    //下移
    void MoveDown(ChessIndex chessIndex)
    {
        switch (selectedChess.type)
        {
            ……
            case ChessType.Rect21:
                chessNextIndex0.rowIndex = chessCurIndex.rowIndex + 1;
                chessNextIndex0.colIndex = chessCurIndex.colIndex;
                chessNextIndex1.rowIndex = chessCurIndex.rowIndex + 1;
                chessNextIndex1.colIndex = chessCurIndex.colIndex + 1;
                MoveRect21(chessCurIndex, chessNextIndex0, chessNextIndex1, movePos);
                Debug.Log("Rect21 MoveDown");
                break;
            ……
        }
    }

    ……

    //移动2*1棋子
    void MoveRect21(ChessIndex chessCurIndex, ChessIndex chessNextIndex0,
                ChessIndex chessNextIndex1, Vector3 movePos)
    {
        //左移或右移
        if (moveDir == MoveDirection.Left || moveDir == MoveDirection.Right)
        {
            if (gameController.GetMapStateInfo(chessNextIndex0.rowIndex,
                chessNextIndex0.colIndex) == 0)
            {
                selectedChess.transform.position += movePos;
                gameController.UpdateMapState(chessCurIndex.rowIndex,
                    chessCurIndex.colIndex, selectedChess.type, moveDir);
            }
        }
        //上移或下移
        else if (moveDir == MoveDirection.Up || moveDir == MoveDirection.Down)
        {
            if (gameController.GetMapStateInfo(chessNextIndex0.rowIndex,
                chessNextIndex0.colIndex) == 0 &&
                gameController.GetMapStateInfo(chessNextIndex1.rowIndex,
                chessNextIndex1.colIndex) == 0)
            {
                selectedChess.transform.position += movePos;
                gameController.UpdateMapState(chessCurIndex.rowIndex,
                    chessCurIndex.colIndex, selectedChess.type, moveDir);
            }
        }
    }
}
```

（3）2×1 棋子的移动效果。返回 Unity 界面，单击"运行"按钮，此时 2×1 棋子已经可以移动，并且棋盘快照也会实时更新，如图 7-41 所示。

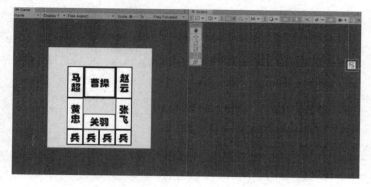

彩色图

图 7-41　2×1 棋子的移动效果

7.4.8　移动棋子——移动 2×2 棋子

2×2 棋子（曹操）的判断点需要考虑棋子自身的横向距离和竖向距离。4 个方向都有判断点。

（1）在 GameController 脚本中实现 2×2 棋子移动后棋盘快照的更新。GameController 脚本中的代码如下：

```csharp
using System.Collections;
using System.Collections.Generic;
using UnityEngine;
using TMPro;

public class GameController : MonoBehaviour
{
    ……

    //更新棋盘快照
    public void UpdateMapState(int chessCurRowIndex, int chessCurColIndex, 
        ChessType type, MoveDirection dir)
    {
        switch (type)
        {
            case ChessType.Rect11:
                ModifyMapStateRect11(chessCurRowIndex, chessCurColIndex, dir);
                break;
            case ChessType.Rect12:
                ModifyMapStateRect12(chessCurRowIndex, chessCurColIndex, dir);
                break;
            case ChessType.Rect21:
                ModifyMapStateRect21(chessCurRowIndex, chessCurColIndex, dir);
                break;
            case ChessType.Rect22:
                ModifyMapStateRect22(chessCurRowIndex, chessCurColIndex, dir);
                break;
        }
    }
```

```
        ShowChessBoardState();    //刷新棋盘快照的显示
    }

    ……

    //2*2 棋子移动后,更新棋盘快照
    void ModifyMapStateRect22(int row, int col, MoveDirection dir)
    {
        switch (dir)
        {
            case MoveDirection.Left:
                mapState[row, col + 1] = 0;
                mapState[row + 1, col + 1] = 0;
                mapState[row, col - 1] = 1;
                mapState[row + 1, col - 1] = 1;
                break;
            case MoveDirection.Right:
                mapState[row, col] = 0;
                mapState[row + 1, col] = 0;
                mapState[row, col + 2] = 1;
                mapState[row + 1, col + 2] = 1;
                break;
            case MoveDirection.Up:
                mapState[row + 1, col] = 0;
                mapState[row + 1, col + 1] = 0;
                mapState[row - 1, col] = 1;
                mapState[row - 1, col + 1] = 1;
                break;
            case MoveDirection.Down:
                mapState[row, col] = 0;
                mapState[row, col + 1] = 0;
                mapState[row + 2, col] = 1;
                mapState[row + 2, col + 1] = 1;
                break;
        }
    }
}
```

(2) 在 MouseEvent 脚本中仿照前面的内容实现 2×2 棋子的移动。MouseEvent 脚本中的代码如下:

```
using System.Collections;
using System.Collections.Generic;
using UnityEngine;

……

public class MouseEvent : MonoBehaviour
{
    ……

    //左移
    void MoveLeft(ChessIndex chessIndex)
    {
```

```csharp
        switch (selectedChess.type)
        {
            ......
            case ChessType.Rect22:
                chessNextIndex0.rowIndex = chessCurIndex.rowIndex;
                chessNextIndex0.colIndex = chessCurIndex.colIndex - 1;
                chessNextIndex1.rowIndex = chessCurIndex.rowIndex + 1;
                chessNextIndex1.colIndex = chessCurIndex.colIndex - 1;
                MoveRect22(chessCurIndex, chessNextIndex0, chessNextIndex1, movePos);
                Debug.Log("Rect22 MoveLeft");
                break;
            ......
        }
    }

    //右移
    void MoveRight(ChessIndex chessIndex)
    {
        switch (selectedChess.type)
        {
            ......
            case ChessType.Rect22:
                chessNextIndex0.rowIndex = chessCurIndex.rowIndex;
                chessNextIndex0.colIndex = chessCurIndex.colIndex + 2;
                chessNextIndex1.rowIndex = chessCurIndex.rowIndex + 1;
                chessNextIndex1.colIndex = chessCurIndex.colIndex + 2;
                MoveRect22(chessCurIndex, chessNextIndex0, chessNextIndex1, movePos);
                Debug.Log("Rect22 MoveRight");
                break;
            ......
        }
    }

    //上移
    void MoveUp(ChessIndex chessIndex)
    {
        switch (selectedChess.type)
        {
            ......
            case ChessType.Rect22:
                chessNextIndex0.rowIndex = chessCurIndex.rowIndex - 1;
                chessNextIndex0.colIndex = chessCurIndex.colIndex;
                chessNextIndex1.rowIndex = chessCurIndex.rowIndex - 1;
                chessNextIndex1.colIndex = chessCurIndex.colIndex + 1;
                MoveRect22(chessCurIndex, chessNextIndex0, chessNextIndex1, movePos);
                Debug.Log("Rect22 MoveUp");
                break;
            ......
        }
    }

    //下移
    void MoveDown(ChessIndex chessIndex)
```

```
{
    switch (selectedChess.type)
    {
    ......
    case ChessType.Rect22:
        chessNextIndex0.rowIndex = chessCurIndex.rowIndex + 2;
        chessNextIndex0.colIndex = chessCurIndex.colIndex;
        chessNextIndex1.rowIndex = chessCurIndex.rowIndex + 2;
        chessNextIndex1.colIndex = chessCurIndex.colIndex + 1;
        MoveRect22(chessCurIndex, chessNextIndex0, chessNextIndex1, movePos);
        Debug.Log("Rect22 MoveDown");
        break;
    ......
    }
}

......

//移动 2*2 棋子
void MoveRect22(ChessIndex chessCurIndex, ChessIndex chessNextIndex0, ChessIndex
                chessNextIndex1, Vector3 movePos)
{
    if (gameController.GetMapStateInfo(chessNextIndex0.rowIndex,
        chessNextIndex0.colIndex) == 0 &&
        gameController.GetMapStateInfo(chessNextIndex1.rowIndex,
        chessNextIndex1.colIndex) == 0)
    {
        selectedChess.transform.position += movePos;
        gameController.UpdateMapState(chessCurIndex.rowIndex,
            chessCurIndex.colIndex, selectedChess.type, moveDir);
    }
}
```

（3）2×2 棋子的移动效果。返回 Unity 界面，单击"运行"按钮，此时 2×2 棋子也可以移动，并且每次移动后棋盘快照都会实时更新，如图 7-42 所示。

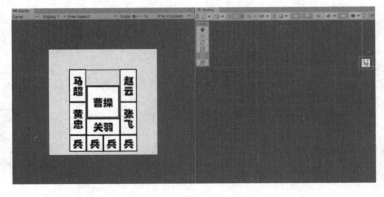

图 7-42　2×2 棋子的移动效果

彩色图

7.4.9　游戏胜利的判断

（1）在 Chess 脚本中添加为 2×2 棋子设定目标位置的函数，以及判断棋子是否到达目标位置的函数。Chess 脚本中的代码如下：

```csharp
using System.Collections;
using System.Collections.Generic;
using UnityEngine;

……

public class Chess : MonoBehaviour
{
……

    public Vector2 targetPosition;    //曹操的目标位置

    ……

    //为 2*2 棋子设定目标位置
    public void SetCaocaoTarget(Vector2 targetPos)
    {
        if (type == ChessType.Rect22)
        {
            targetPosition = targetPos;
        }
    }

    //判断是否到达目标位置
    public bool IsReachTargetPos()
    {
        return (((int)targetPosition.x == (int)transform.position.x) &&
                ((int)targetPosition.y == (int)transform.position.y));
    }
}
```

（2）找到 GameController 脚本，在初始化棋子时调用 SetCaocaoTarget() 函数。GameController 脚本中的代码如下：

```csharp
using System.Collections;
using System.Collections.Generic;
using UnityEngine;
using TMPro;

public class GameController : MonoBehaviour
{
    ……

    //初始化棋盘盘面
    void InitGame()
    {
        ……
        chessObjs[9].SetCaocaoTarget(new Vector2(2, 2));
    }
}
```

（3）因为只有在移动 2×2 棋子后才需要判断游戏是否胜利，所以在移动 2×2 棋子的函数中加一个游戏是否胜利的判断。MouseEvent 脚本中的代码如下：

```csharp
using System.Collections;
```

```csharp
using System.Collections.Generic;
using UnityEngine;

……

public class MouseEvent : MonoBehaviour
{
    ……

    //移动 2*2 棋子
    void MoveRect22(ChessIndex chessCurIndex, ChessIndex chessNextIndex0,
    ChessIndex chessNextIndex1, Vector3 movePos)
    {
        if (gameController.GetMapStateInfo(chessNextIndex0.rowIndex,
            chessNextIndex0.colIndex) == 0 &&
            gameController.GetMapStateInfo(chessNextIndex1.rowIndex,
            chessNextIndex1.colIndex) == 0)
        {
            selectedChess.transform.position += movePos;
            gameController.UpdateMapState(chessCurIndex.rowIndex,
                chessCurIndex.colIndex, selectedChess.type, moveDir);
            GameOver();
        }
    }

    //游戏胜利的判断
    void GameOver()
    {
        if (selectedChess.IsReachTargetPos())
        {
            Debug.Log("GameOver");
        }
    }
}
```

（4）设置游戏胜利画面的图片。在主菜单栏中选择 GameObject→UI→Image 命令，新建一个 Image 类型的 UI 对象，并且把 Sprites 文件夹下的 gameOver 图片赋给 Image 组件的 Source Image，如图 7-43 所示。

（5）摆放游戏胜利的 UI 图片。将 gameOver 图片居中对齐到屏幕中央，并调整至合适的大小，如图 7-44 所示。

图 7-43　设置游戏胜利画面的图片

图 7-44　摆放游戏胜利的 UI 图片

（6）在设置好游戏胜利的图片后，把它的激活状态变为否，如图 7-45 所示。

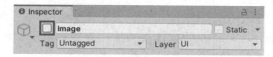

图 7-45　把游戏胜利的图片的激活状态变为否

（7）在游戏胜利的判断函数中增加一条语句，当判断游戏胜利时，将游戏胜利的图片的激活状态改变为 true。MouseEvent 脚本中的代码如下：

```
using System.Collections;
using System.Collections.Generic;
using UnityEngine;
using UnityEngine.UI;

public class MouseEvent : MonoBehaviour
{
    ……
    public Image gameOverImage;   //游戏胜利的图片
    ……
    //游戏胜利的判断
    void GameOver()
    {
        if (selectedChess.IsReachTargetPos())
        {
            gameOverImage.gameObject.SetActive(true);
            Debug.Log("GameOver");
        }
    }
}
```

（8）返回 Unity 界面，将创建好的游戏胜利的图片的 Image 组件赋值给 MouseEvent 脚本中的 Game Over Image 参数，如图 7-46 所示。

（9）单击"运行"按钮，如果没有出错，则游戏胜利的画面如图 7-47 所示。

图 7-46　为 Game Over Image 参数赋值

图 7-47　游戏胜利的画面

7.5 本章小结

本章主要介绍了《华容道》游戏的设计与制作，其中重点在于棋子的数据结构表达、如何对不同类型的棋子进行定位，以及如何移动棋子等。

7.6 练习题

1．目前，《华容道》游戏的地图快照是由整型二维数组表示的，棋子和墙体所在区域编号为1，空地编号为0。由于空地数量较少，因此可以考虑另一种表达方式，即只记录空地位置坐标集合，以代替整型二维数组，此想法是否可行？试分析其相比原结构的优点和缺点，以及后续实现时需要注意的问题。

2．在《华容道》游戏的交互方式上，目前采用的是鼠标拖动的方式，分别记录按下和松开鼠标按键的位置坐标，并根据鼠标指针的位移确定移动方向。现考虑只用键盘上的方向键来实现棋子的选择和移动，该想法是否可行？请说明理由。

3．考虑《华容道》游戏的一种简单的自动求解算法（广度优先搜索）：将不同的棋盘盘面状态看作不同的节点，从初始棋盘盘面状态开始，遍历当前状态下所有合法的走法，并且将当前状态作为所有下一个可能状态的父节点，如此进行直至找到2×2棋子到达目标位置的状态为止。假设此过程中不会出现环（每次加入新状态前均需要确保与先前所有状态无重复），如此即可构成一棵庞大的状态树。现假设每个状态的下一个合法状态平均有 k 种，且成功结束游戏平均需要 n 步，请问该求解算法的时间复杂度为多少？

第 8 章

连连看

8.1 游戏简介

《连连看》是由黄兴武创作的一款 PC 端益智类游戏，于 2001 年成型并迅速传播开来，被选为 2010 年上海世博会的官方推荐游戏。这款游戏的规则简单易懂，只需要将相同的两张图片用 3 根以内（包含 3 根）的直线连接起来即可消除。另外，这款游戏的画面可爱清晰，适合细心的玩家，丰富的道具和公共模式增加了竞争性。多样化的地图使玩家在各个游戏水平上都能找到挑战目标，从而保持游戏的新鲜感。《连连看》游戏的画面如图 8-1 所示。

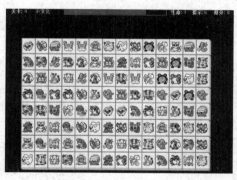

图 8-1 《连连看》游戏的画面

8.2 游戏规则

将相同的两张图片用 3 根以内（包含 3 根）的直线连在一起便可消除，将全地图图片消除完即可获胜。《连连看》游戏最关键的玩法在于可消除的 3 种情况：直连（1 根直线），如图 8-2 所示；二连（2 根直线），如图 8-3 所示；三连（3 根直线），如图 8-4 所示。

图 8-2 直连　　　　　　图 8-3 二连　　　　　　图 8-4 三连

8.3 游戏开发核心思路

8.3.1 编写 Tile 类

（1）Tile 类的功能：使用整型记录图片在场景中的设置 (x,y)，以及图片的 ID。假设图片的 x 坐标为 PosX，y 坐标为 PosY；图片类型用 ID 表示。

（2）将 Tile 脚本挂载到一个添加了 SpriteRenderer 的游戏对象预制体上。

8.3.2 初始化地图

（1）使用一维数组保存游戏所需的 Tile 图片，如图 8-5 所示。

图 8-5 使用一维数组保存游戏所需的 Tile 图片

（2）创建一个二维数组，用于保存图片类型的 ID，称为 ID 数组。随机获取图片数组的索引号（该 Tile 图片的 ID），并将其保存到一个二维数组中，在下一个相邻位置上保存相同的 ID，这样可以保证 Tile 图片两两配对。图 8-6 所示为相邻元素 ID 相同的 ID 数组。

（3）将 ID 数组打乱。将图片打乱，ID 数组打乱后的效果如图 8-7 所示。

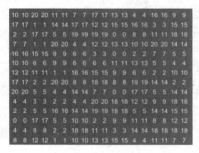

图 8-6 相邻元素 ID 相同的 ID 数组

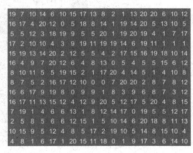

图 8-7 ID 数组打乱后的效果

（4）创建一个地图快照二维数组，并加上外围墙面。图 8-8 所示为加上外围墙面的地图快照效果。

（5）创建和绘制 Tile 对象。根据 snapshot 数组创建和绘制 Tile 对象，如图 8-9 所示。

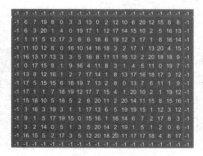

图 8-8 加上外围墙面的地图快照效果

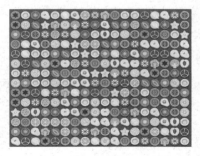

图 8-9 创建和绘制 Tile 对象

8.3.3 消除操作

消除操作的流程图如图 8-10 所示。

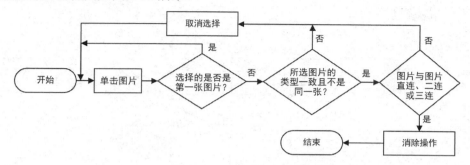

图 8-10 消除操作的流程图

1．消除操作——选择两张图片

（1）从摄像机发射射线。
（2）如果射线与某张图片相交，则表示选中了当前的图片。
（3）如果选择的是第一张图片，则保存该对象。
（4）如果选择的是第二张图片，则保存该对象。
（5）判断是否是同一张图片，若为同一张图片，则位置相同，即判断两张图片对应的 x 和 y 是否完全一致。
（6）判断选择的图片的类型是否一致，即判断两张图片的类型 ID 是否相同。
（7）如果类型 ID 相同，则进行可消除判断。

选择两张图片的流程图如图 8-11 所示。

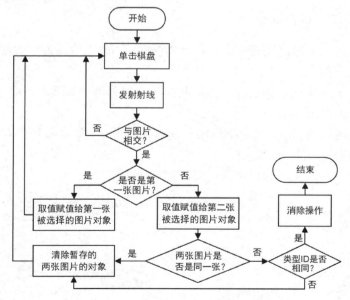

图 8-11 选择两张图片的流程图

2．消除操作——连接判断框架

连接判断框架的流程图如图 8-12 所示。

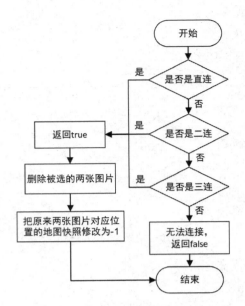

图 8-12　连接判断框架的流程图

3. 消除操作——直连的判断

直连：两张图片可以被 1 根直线连接。直连的判断流程如下。

（1）获取两张图片的坐标，用坐标遍历判断两张图片之间是否有障碍物。

（2）设选择的第一张图片和第二张图片的坐标分别为（x_1,y_1）和（x_2,y_2）。

（3）直连情况可以分为横向直连和纵向直连：横向直连的 y_1 和 y_2 相等。纵向直连的 x_1 和 x_2 相等，如图 8-13 所示。

4. 消除操作——二连的判断

二连：两张图片被 2 根直线连接，且有一个折角，因此也被称为一折连接，如图 8-14 所示。二连的判断流程如下。

（1）获取两张被选择图片的坐标，并以两张图片为对角顶点形成矩形。设两张图片的位置坐标分别为（x_1,y_1）和（x_2,y_2）。

（2）以第一张图片为基准，分别从纵向和横向寻找拐点，如果与某个方向上的拐点直连，则判断该拐点是否与第二张图片直连。

（3）如果两张图片与某个拐点都是直连，则二连成立。

图 8-13　两张图片以 1 根直线连接　　　　　图 8-14　二连情况

5．消除操作——三连的判断

三连：两张图片可以被 3 根相连的直线连接，因为有 2 个折角，所以又被称为二折。三连的判断流程如下。

（1）以第一张被选择的图片为中心，通过右探、左探、上探和下探寻找红色点（见图 8-15～图 8-18），如果遍历过程中遇到有图片的地方，则说明连接过程中有其他图片阻挡连接操作，应停止当前遍历。

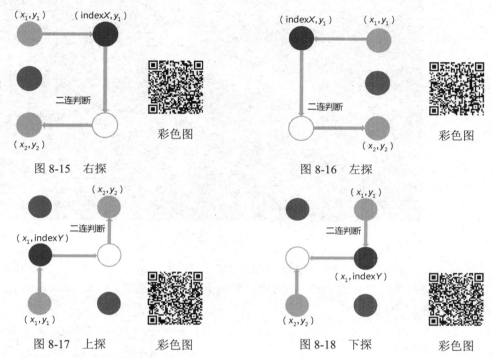

（2）红色点的条件是能够与第二张被选择的图片满足二连判断。

（3）绿色点为已经选择的图片，土色点为有其他图片。

8.3.4 绘制连接线

把连接线绘制出来可以为玩家呈现一定的游戏反馈，使游戏更加生动。直连为 2 个端点，如图 8-19 所示；二连为 3 个端点，如图 8-20 所示；三连为 4 个端点，如图 8-21 所示。

图 8-19　直连

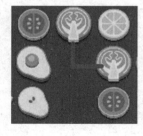
图 8-20　二连

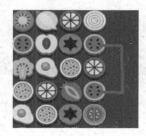
图 8-21　三连

8.4 游戏实现

8.4.1 导入资源

（1）创建项目。如图 8-22 所示，选择 2D 选项，创建一个名称为 Link Up 的项目。

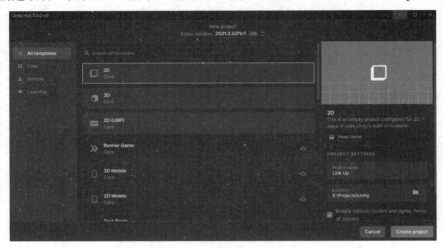

图 8-22　创建项目

（2）导入图片素材。将下载好的图片素材保存到 Sprites 文件夹中，如图 8-23 所示。

（3）切割图片。单击 LinkUp 图片，将 Sprite Mode 设置为 Multiple，单击 Sprite Editor 按钮，如图 8-24 所示。

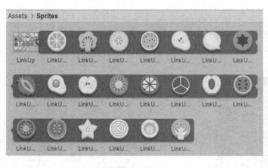

图 8-23　导入图片素材

图 8-24　切割图片

① 设置行和列。单击 Slice 下拉按钮，将 Type 设置为 Grid By Cell Count，即按行和列的数量切割。将 Column & Row 设置为 6 和 4，单击 Slice 按钮，如图 8-25 所示。

② 单击 Apply 按钮，完成切割后的效果如图 8-26 所示。

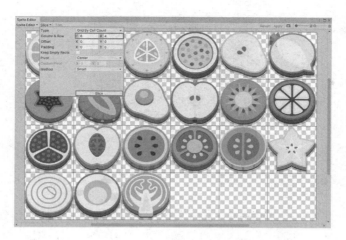

图 8-25　设置行和列

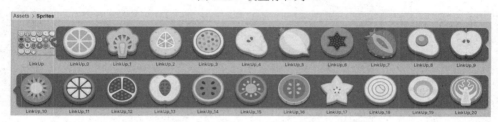

图 8-26　完成切割后的效果

（4）设置摄像机。将摄像机的位置设置为（10,8,-10），Size 设置为 9，如图 8-27 所示。

① 设置图片大小。将每张图片设置成一个 Unity 单位，将 Pixels Per Unit 设置为 128，如图 8-28 所示。

图 8-27　设置摄像机

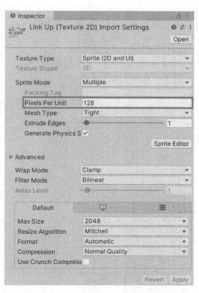

图 8-28　设置图片大小

② 单击 Apply 按钮，将图片拖到 Scene 窗口中观察，若每张图片占一个 Unity 单位则设置成功，如图 8-29 所示。

图 8-29 图片大小设置成功

8.4.2 编写 Tile 类

（1）在 Scripts 文件夹下新建一个 C#脚本并将其命名为 Tile，打开该脚本。

（2）编写 Tile 脚本，即在该脚本中定义图片的一些属性。Tile 脚本中的代码如下：

```csharp
using System.Collections;
using System.Collections.Generic;
using UnityEngine;

public class Tile : MonoBehaviour
{
    public int posX{get; private set;}          //图片在数组中的X位置
    public int posY{ get; private set; }        //图片在数组中的Y位置
    public int ID { get; private set; }         //图片的贴图

    //设置初始值
    public void SetInitValue(int posX,int posY,int ID)
    {
        this.posX = posX;
        this.posY = posY;
        this.ID = ID;
    }
}
```

（3）返回 Unity 界面，在主菜单栏中选择 GameObject→Creat Empty 命令，创建一个新的游戏对象。按 F2 键将新创建的游戏对象重命名为 Tile。单击 Tile 对象，在 Inspector 窗口中单击最下方的 Add Component 按钮，搜索 Sprite Renderer。将 Tile 脚本拖到 Inspector 窗口中，创建一个新的游戏对象，如图 8-30 所示，加入 Sprite Renderer 组件并添加 Tile 脚本，如图 8-31 所示。

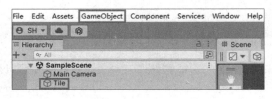

图 8-30 创建一个新的游戏对象

图 8-31 加入 Sprite Renderer 组件并添加 Tile 脚本

（4）设置预制体。将 Tile 对象拖到 Prefabs 文件夹中，如图 8-32 所示。

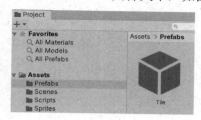

图 8-32　设置预制体

8.4.3　初始化地图

（1）在 Scripts 文件夹中新建一个 C#脚本并将其命名为 MapController。MapController 脚本中的代码如下：

```csharp
using System.Collections;
using System.Collections.Generic;
using UnityEngine;

public class MapController : MonoBehaviour
{
    public int rowNum = 14;                //Tile 的总行数
    public int colNum = 18;                //Tile 的总列数
    public Sprite[] tileSprites;           //Tile 的图片数组
    private int[,] IDMap;                  //ID 数组
    private int[,] snapshotMap;            //快照数组
    public TextMesh[,] texts;              //用于显示地图信息
    public TextMesh textPrefab;
    //为了保持 Hierarchy 窗口的简洁，用一个父物体挂载 texts 对象
    public Transform textRoot;
    public Vector2 textOffset = new Vector2(1.5f, 1.5f);

    void Start()
    {
        CreateIDMap();                     //创建 ID 数组
    }

    void CreateIDMap()
    {
        IDMap = new int[rowNum, colNum];
        for (int rowIndex = 0; rowIndex < rowNum; rowIndex++)
        {
            for (int colIndex = 0; colIndex < colNum; colIndex = colIndex + 2)
            {   //注意列号的递增
                //随机获得图片数组的索引
                int tileIndex = Random.Range(0, tileSprites.Length);
                IDMap[rowIndex, colIndex] = tileIndex;     //把索引存储到 ID 数组中
                //下一个相邻位置保存相同的索引
                IDMap[rowIndex, colIndex + 1] = tileIndex;
                //在场景中显示当前的 IDMap 数值
                ShowMapInfo(rowIndex, colIndex, IDMap[rowIndex, colIndex]);
```

```
                ShowMapInfo(rowIndex, colIndex + 1, IDMap[rowIndex, colIndex + 1]);
        }//colIndex
    }//rowIndex
}

void ShowMapInfo(int row, int col, int value)
{
    if (texts == null)
    {
        //为了能够与地图快照对应，这里将 rowNum 和 colNum 分别加 2
        texts = new TextMesh[rowNum + 2, colNum + 2];
    }//texts == null

    if (texts[row, col] == null)
    {
        TextMesh tempText = Instantiate(textPrefab, new Vector2(col + textOffset.x,
                                        row+textOffset.y), Quaternion.identity);
        texts[row, col] = tempText;
        tempText.transform.SetParent(textRoot);
    }
    texts[row, col].text = value.ToString();
}
```

（2）返回 Unity 界面，在主菜单栏中选择 GameObject→3D Object→3D Text 命令，新建一个 Text 并将其命名为 MapInfoText，将 Text 组件中的内容删除，将对齐方式设置为 Middle Center 和 Center。图 8-33 所示为设置 MapInfoText。将 MapInfoText 拖到 Prefabs 文件夹中设置预制体，如图 8-34 所示。

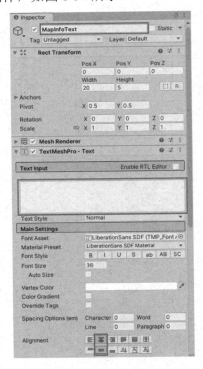

图 8-33　设置 MapInfoText

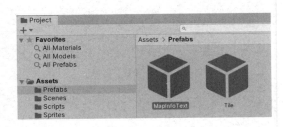

图 8-34　设置 MapInfoText 预制体

（3）创建一个名称为 MapController 的空对象，并将 MapController 脚本拖到该对象下。将 Position 归零，并把刚做的 MapInfoText 预制体拖到 MapController 对象的 MapController 脚本的 Text Prefab 参数上，把 MapController 对象拖到 MapController 脚本的 Text Root 参数上，如图 8-35 所示。为 Tile 分配的元素为 21 个，如图 8-36 所示。

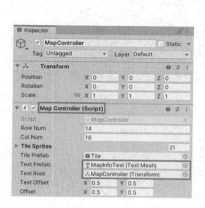

图 8-35　拖动并配置 MapController 脚本

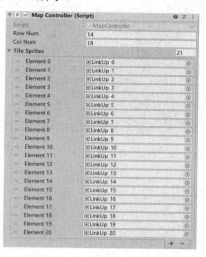

图 8-36　为 Tile 分配元素

单击"运行"按钮，可以看到 ID 数组效果，如图 8-37 所示。

（4）返回 MapController 脚本，编写打乱图片的代码。MapController 脚本中的代码如下：

```
using System.Collections;
using System.Collections.Generic;
using UnityEngine;

public class MapController : MonoBehaviour
{
    ……
    void Start()
    {
        CreateIDMap();          //创建 ID 数组
        Shuffle();              //打乱图片
    }
    ……
    void Shuffle()
    {
        for (int rowIndex = 0; rowIndex < rowNum; rowIndex++)
        {
            for (int colIndex = 0; colIndex < colNum; colIndex++)
            {
                int temp = IDMap[rowIndex, colIndex];
                int randomRow = Random.Range(0, rowNum);
                int randomCol = Random.Range(0, colNum);
                IDMap[rowIndex, colIndex] = IDMap[randomRow, randomCol];
                IDMap[randomRow, randomCol] = temp;
                ShowMapInfo(rowIndex, colIndex, IDMap[rowIndex, colIndex]);
            }//colIndex
        }//row
    }
}
```

（5）返回 Unity 界面，运行后发现索引号被打乱了，实现打乱图片的功能。打乱图片的效果如图 8-38 所示。

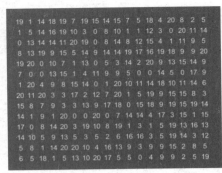

图 8-37　ID 数组效果　　　　　　　　图 8-38　打乱图片的效果

（6）返回脚本编辑器，创建一个地图快照的二维数组。MapController 脚本中的代码如下：

```
using System.Collections;
using System.Collections.Generic;
using UnityEngine;

public class MapController : MonoBehaviour
{
    ……
    void Start()
    {
        CreateSnapshotMap();      //创建地图快照
    }

    ……

    void CreateSnapshotMap()
    {
        //为快照数组分配内存空间
        snapshotMap = new int[rowNum + 2, colNum + 2];
        for (int rowIndex = 0; rowIndex < rowNum + 2; rowIndex++)
        {       //快照的行数比 ID 数组的行数多两行
            for (int colIndex = 0; colIndex < colNum + 2; colIndex++)
            {   //快照的列数比 ID 数组的行数多两列
                //设置墙面
                if (rowIndex == 0 || rowIndex == rowNum + 1 || colIndex == 0 ||
                  colIndex == colNum + 1)
                {
                    snapshotMap[rowIndex, colIndex] = -1;   //墙面 ID 设置为-1
                }
                else
                {
                    snapshotMap[rowIndex, colIndex] = IDMap[rowIndex - 1,
                                                           colIndex - 1];
                    //把对应的 ID 赋值给地图快照，但需要注意索引号
                }
                ShowMapInfo(rowIndex, colIndex, snapshotMap[rowIndex,
                                                   colIndex]);

            }//colIndex
```

```
        }//rowIndex
    }
}
```

（7）为游戏场景添加墙面效果。返回 Unity 界面，单击"运行"按钮，可以发现第一行和最后一行，以及第一列和最后一列都被附上墙面了，如图 8-39 所示。

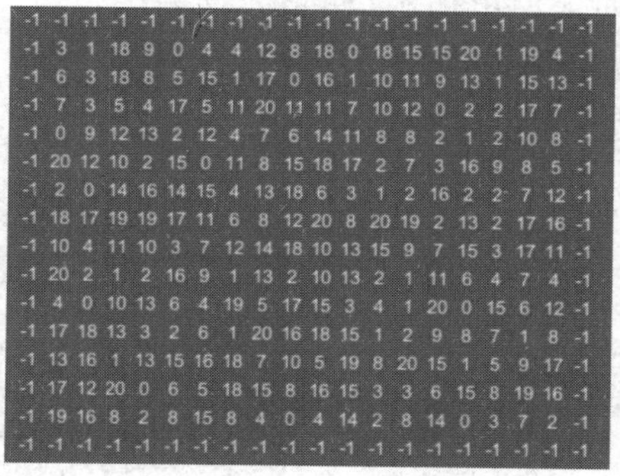

图 8-39　为游戏场景添加墙面效果

（8）返回脚本编辑器，实现根据地图快照显示图片的功能。MapController 脚本中的代码如下：

```
using System.Collections;
using System.Collections.Generic;
using UnityEngine;

public class MapController : MonoBehaviour
{
    public GameObject tilePrefab;                   //Tile 预制体
    public Vector2 tileOffset = new Vector2(0.5f, 0.5f);
    ……

    void Start()
    {
        PaintTiles();                               //绘制场景
    }
    ……
    void PaintTiles()
    {
        GameObject tempTile;
        for (int tilePosY = 1; tilePosY < rowNum + 1; tilePosY++)
        {   //Y 坐标
            for (int tilePosX = 1; tilePosX < colNum + 1; tilePosX++)
            {   //X 坐标
                tempTile = Instantiate(tilePrefab) as GameObject;
                //根据实例化后的 Tile 类型在地图快照中的索引号+偏移量，为当前生成的 Tile 赋予位置
                tempTile.transform.position = new Vector3(tilePosX + tileOffset.x,
                                                tilePosY + tileOffset.y,0);
                //为了方便管理，为每个 Tile 起一个以当前位置坐标为数值的名称
                tempTile.name = "Tile" + tilePosX + "_" + tilePosY;
                //将在地图快照中取出的 tileID，作为取出 tileSprites 数组中的图片的索引号
                tempTile.GetComponent<SpriteRenderer>().sprite=tileSprites[
                                        snapshotMap[tilePosY,tilePosX]];
                tempTile.GetComponent<Tile>().SetInitValue(tilePosX,tilePosY,
                                        snapshotMap[tilePosY, tilePosX]);
```

```
            }//tilePosX
        }//tilePosY
    }
}
```

（9）返回 Unity 界面，将每张图片拖到 Tile Sprites 数组中，把 Prefabs 文件夹中的 Tile 预制体赋给 Tile Prefab 参数，如图 8-40 所示。单击"运行"按钮，生成的地图如图 8-41 所示。

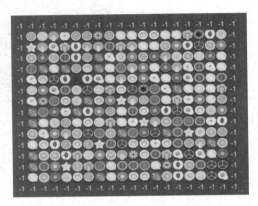

图 8-40　把每个图案赋给 Tile Sprites 数组　　　　图 8-41　生成的地图

8.4.4　选择 Tile 对象

（1）在 Scripts 文件夹中新建一个 C#脚本并将其命名为 Link。Link 脚本中的代码如下：

```
using System.Collections;
using System.Collections.Generic;
using UnityEngine;

public enum LinkType{Direct,OneCorner,TwoCorner}

public class Link : MonoBehaviour
{
    Ray ray;
    RaycastHit2D hit = new RaycastHit2D();
    public Camera mainCamera;
    private Tile firSelectedTile,secSelectedTile;

    void SelectTile(){
        if(Input.GetButtonDown("Fire1")){
            //从屏幕上鼠标指针的位置发射一条射线
            ray = mainCamera.ScreenPointToRay(Input.mousePosition);
            hit = Physics2D.Raycast(ray.origin,ray.direction);
            if(hit.collider != null){
                //如果第一张图片的变量为空，则表示第一张图片未被选择
                if(firSelectedTile == null){
                    //把选择的图片的指针取出并赋值给 firstSelectedTile
                    firSelectedTile = hit.transform.gameObject.GetComponent<Tile>();
                    //改变当前选择的图片的颜色，表示已经选上
                    firSelectedTile.GetComponent<SpriteRenderer>().color = Color.red;
                }else{
```

```
                    //如果已经选择第一张图片,
                    //则把被选择的图片的指针赋值给第二张图片的变量
                    secSelectedTile = hit.transform.gameObject.GetComponent<Tile>();
                    //改变当前选择的图片的颜色,表示已经选上
                    secSelectedTile.GetComponent<SpriteRenderer>().color = Color.red;
                }
            }//hit.collider
        }//GetButtonDown
    }
}
```

图 8-42 添加碰撞盒

（2）返回 Unity 界面，为 Tile 对象加上碰撞盒，单击 Add Component 按钮，搜索 Box Collider 2D 并添加组件，如图 8-42 所示。将 Link 脚本赋给 MapController（见图 8-43），并把 Main Camera 拖到相应的位置。单击"运行"按钮，发现单击每张图片时图案会变成红色。图 8-44 所示为添加碰撞盒后图案被选择的效果。

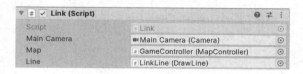

图 8-43　将 Link 脚本赋给 MapController

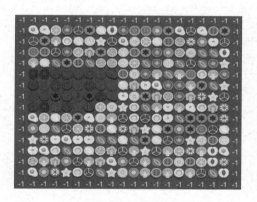

图 8-44　添加碰撞盒后图案被选择的效果

（3）实现清除选择的功能。返回脚本编辑器，实现清除选择的功能。如果不能相连或 ID 不同，则清除选择；如果选择的是同一张图片，则清除选择。Link 脚本中的代码如下：

```
using System.Collections;
using System.Collections.Generic;
using UnityEngine;

public class Link : MonoBehaviour
{
    Ray ray;
    RaycastHit2D hit = new RaycastHit2D();
    public Camera mainCamera;
    private Tile firSelectedTile,secSelectedTile;
    void SelectTile(){
```

```csharp
            //使用按键别名实现单击鼠标左键
            if(Input.GetButtonDown("Fire1")){
                //从屏幕上鼠标指针的位置发射一条射线
                ray = mainCamera.ScreenPointToRay(Input.mousePosition);
                //检测该射线是否与场景中的碰撞盒相交，若相交，则将相关数据返回给 hit
                hit = Physics2D.Raycast(ray.origin,ray.direction);
                if(hit.collider != null){
                    //如果第 1 张图片的变量为空，则表示第 1 张图片未被选择
                    if(firSelectedTile == null){
                        //把选择的图片的指针取出并赋值给 firstSelectedTile
                        firSelectedTile = hit.transform.gameObject.GetComponent<Tile>();
                        //改变当前选择的图片的颜色，表示已经选上
                        firSelectedTile.GetComponent<SpriteRenderer>().color = Color.red;
                    }else{
                        //如果已经选择第一张图片，则把被选择的牌的指针赋值给第二张图片的变量
                        secSelectedTile = hit.transform.gameObject.GetComponent<Tile>();
                        //改变当前选择的图片的颜色，表示已经选上
                        secSelectedTile.GetComponent<SpriteRenderer>().color =Color.red;
                        IsTheSame();

                    }
                }//hit.collider
            }//GetButtonDown
    }
    void IsTheSame(){
        //1.如果选择的图片不是同一张
        if(firSelectedTile.transform.position != secSelectedTile.transform.position){
            //2.如果选择的图片的 ID 相同
            if(firSelectedTile.ID == secSelectedTile.ID){
                //3.判断是否可以连接，如果 ID 相同，则进行相连判断，否则清除已经被选择的图片的暂存变量
                if(false){

                }else{
                    //如果不可相连，则清除已经被选择的图片的暂存变量
                    ClearSelectedInfo();
                }
            }else{
                //如果 ID 不相等，则清除已经被选择的图片的暂存变量
                ClearSelectedInfo();
            }
        }else{
            //如果选择的是同一张图片，则清除已经被选择的图片的暂存变量
            ClearSelectedInfo();
        }
    }
    void ClearSelectedInfo(){
        //把选择的图片的颜色改回白色
        firSelectedTile.GetComponent<SpriteRenderer>().color = Color.white;
        secSelectedTile.GetComponent<SpriteRenderer>().color = Color.white;
        //清空图片的暂存变量
        firSelectedTile = null;
        secSelectedTile = null;
    }
}
```

8.4.5 连接判断——直连、二连、三连

（1）返回 IsTheSame()函数，编写清除操作，以及 IsLink()函数框架。Link 脚本中的代码如下：

```
void IsTheSame(){
    //1.如果选择的图片不是同一张
    if(firSelectedTile.transform.position != secSelectedTile.transform.position){
        //2.如果选择的图片的 ID 相同
        if(firSelectedTile.ID == secSelectedTile.ID){
            //3.判断是否可以连接，如果 ID 相同，则进行相连判断，否则清除已经被选择的图片的暂存变量
            if(IsLink(firSelectedTile.PosX,firSelectedTile.PosY,secSelectedTile.PosX,
                secSelectedTile.PosY)){
            }else{
                //如果不可相连，则清除已经被选择的图片的暂存变量
                ClearSelectedInfo();
            }
        }else{
            //如果 ID 不相同，则清除已经被选择的图片的暂存变量
            ClearSelectedInfo();
        }
    }else{
        //如果选择的是同一张图片，则清除已经被选择的图片的暂存变量
        ClearSelectedInfo();
    }
}
bool IsLink(int x1,int y1,int x2,int y2){
    //直连
    //二连
    //三连
    return false;
}
```

（2）在函数框架中实现具体的相连判断。Link 脚本中的代码如下：

```
using System.Collections;
using System.Collections.Generic;
using UnityEngine;

public class Link : MonoBehaviour
{
    ......
    public MapController snapshotMap;
    ......
    bool IsLink(int x1,int y1,int x2,int y2){
        //直连
        //垂直方向
        if(x1 == x2){
            if(Y_LinkDetection(x1,y1,y2)){
                return true;
            }
        }
        //水平方向
        if(y1 == y2){
            if(X_LinkDetection(x1,x2,y1)){
```

```
            return true;
        }
    }
    //二连
    //三连
    return false;
}
bool Y_LinkDetection(int x1,int y1,int y2){
    if(y1 > y2){      //为了统一计算,此时通过交换保证 y2 > y1
        int exchange = y1;
        y1 = y2;
        y2 = exchange;
    }

    //开始纵向遍历,查看两张图片之间是否有其他图片
    for(int indexY = y1+1; indexY <= y2;indexY++){
        //若相邻,则返回 true,退出函数
        if(indexY == y2){return true;}
        //-1 表示当前位置为空,没有其他图片。如果两张图片之间有其他图片,则它们不相连,直接退出循环
        if(snapshotMap.GetMapItem(indexY,x1) != -1){break;}  //退出循环
    }
    return false;
}

bool X_LinkDetection(int x1,int x2,int y2){
    if(x1 > x2){
        int exchange = x1;
        x1 = x2;
        x2 = exchange;
    }

    for(int indexX = x1+1;indexX <= x2;indexX++){
        //若相邻,则返回 true,退出函数
        if(indexX == x2){return true;}
        //-1 表示当前位置为空,没有其他图片。如果两张图片之间有其他图片,则它们不相连,直接退出循环
        if(snapshotMap.GetMapItem(y2,indexX) != -1){break;}//退出循环
    }
    return false;
}
}
```

（3）在 MapController 脚本中实现 GetMapItem()函数。MapController 脚本需要添加的代码如下：

```
public int GetMapItem(int row,int col){
    return snapshotMap[row,col];
}
```

编写方法清除相连的两个 Tile，并更新快照。Link 脚本需要添加的代码如下：

```
void IsTheSame(){
    //1.如果选择的图片不是同一张
    if(firSelectedTile.transform.position != secSelectedTile.transform.position){
        //2.如果选择的图片的 ID 相同
        if(firSelectedTile.ID == secSelectedTile.ID){
            //3.判断是否可以连接,如果 ID 相同,则进行相连判断,否则清除已经被选择的图片的暂存变量
```

```
                if(IsLink(firSelectedTile.PosX,firSelectedTile.PosY,secSelectedTile.PosX,
                secSelectedTile.PosY)){
                    DestroySameItem(firSelectedTile.PosX,firSelectedTile.PosY,
                    secSelectedTile.PosX,secSelectedTile.PosY);
                }else{
                    //如果不可相连，则清除已经被选择的图片的暂存变量
                    ClearSelectedInfo();
                }
            }else{
                //如果ID不相同，则清除已经被选择的图片的暂存变量
                ClearSelectedInfo();
            }
    }else{
        //如果选择的是同一张图片，则清除已经被选择的图片的暂存变量
        ClearSelectedInfo();
    }
}
void DestroySameItem(int x1,int y1, int x2, int y2){
    Destroy(firSelectedTile.gameObject);
    Destroy(secSelectedTile.gameObject);
    snapshotMap.SetMapItem(y1,x1,-1);
    snapshotMap.SetMapItem(y2,x2,-1);
    firSelectedTile = null;
    secSelectedTile = null;
}
```

MapController 脚本需要添加的代码如下：

```
public void SetMapItem(int row,int col,int value){
    snapshotMap[row,col] = value;
    ShowMapInfo(row,col,value);
}
```

（4）将 MapController 脚本赋给 Snapshot Map 参数。返回 Unity 界面，将 MapController 脚本赋给 Snapshot Map 参数并运行，如图 8-45 所示。运行游戏，发现可以消除直连的两张图片。

图 8-45　将 MapController 脚本赋给 Snapshot Map 参数

（5）实现二连功能。Link 脚本需要添加的代码如下：

```
bool IsLink(int x1,int y1,int x2,int y2){
    ......
    //二连
    if(OneCornerLinkDetection(x1,y1,x2,y2)){
        return true;
    }
    return false;
}
bool OneCornerLinkDetection(int x1, int y1, int x2, int y2){
    //垂直拐角，没有其他图片
    //  (x1,y1)  *********
```

```
//        *********
// (x1,y2)+ ********(x2,y2)
if(snapshotMap.GetMapItem(y2,x1) == -1){
    if(X_LinkDetection(x1,x2,y2) && Y_LinkDetection(x1,y1,y2)){
        return true;
    }
}
//水平拐角,没有其他图片
// (x1,y1)********+(x2,y1)
//        *********
//        *********(x2,y2)
if(snapshotMap.GetMapItem(y1,x2) == -1){
    if(X_LinkDetection(x1,x2,y1) && Y_LinkDetection(x2,y1,y2)){
        return true;
    }
}
return false;
}
```

(6) 实现三连功能。MapController 脚本需要添加的代码如下:

```
public int GetMapRowNum(){
    return snapshotMap.GetLength(0);
}
public int GetMapColNum(){
    return snapshotMap.GetLength(1);
}
```

Link 脚本需要添加的代码如下:

```
bool IsLink(int x1,int y1,int x2,int y2){
……
    //////三连
    if(TwoCornerLinkDetection(x1,y1,x2,y2)){
      return true;
    }
return false;
}
bool TwoCornerLinkDetection(int x1, int y1, int x2, int y2){
    ////右探
    for(int indexX = x1+1; indexX < snapshotMap.GetMapColNum();indexX++){
        //下一个位置没有其他图片
        if(snapshotMap.GetMapItem(y1,indexX) == -1){
            //用当前位置与第二张图片进行二连判断
            if(OneCornerLinkDetection(indexX,y1,x2,y2)){
                return true;
            }
        }
        //如果下一个位置有其他图片,则退出右探遍历
        if(snapshotMap.GetMapItem(y1,indexX) != -1){break;}
    }
    ////左探
    for(int indexX = x1-1;indexX >= 0;indexX--){
        if(snapshotMap.GetMapItem(y1,indexX) == -1){
            if(OneCornerLinkDetection(indexX,y1,x2,y2)){
                return true;
            }
```

```
            }
            if(snapshotMap.GetMapItem(y1,indexX) != -1){break;}
        }
        ////上探
        for(int indexY = y1+1;indexY < snapshotMap.GetMapRowNum();indexY++){
            if(snapshotMap.GetMapItem(indexY,x1) == -1){
                if(OneCornerLinkDetection(x1,indexY,x2,y2)){
                    return true;
                }
            }
            if(snapshotMap.GetMapItem(indexY,x1) != -1){break;}
        }
        ////下探
        for(int indexY = y1-1;indexY >= 0;indexY--){
            if(snapshotMap.GetMapItem(indexY,x1) == -1){
                if(OneCornerLinkDetection(x1,indexY,x2,y2)){
                    return true;
                }
            }
            if(snapshotMap.GetMapItem(indexY,x1) != -1){break;}
        }
        return false;
    }
```

8.4.6 连接线的绘制

（1）返回 Unity 界面，在主菜单中选择 GameObject→Effect→Line 命令，新建一个 Line 对象并将其命名为 LinkLine（见图 8-46），修改连接线的颜色（见图 8-47）。

图 8-46 新建连接线对象

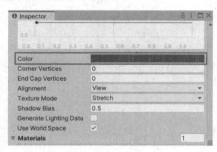

图 8-47 改变连接线的颜色

（2）新建 LineCreator 脚本。LineCreator 脚本中的代码如下：

```
using System.Collections;
using System.Collections.Generic;
using UnityEngine;

public class LineCreator : MonoBehaviour
{
    public LineRenderer line;

    public void DrawLinkLine(Transform firstTile,Transform secondTile,LinkType 
        linkType,Vector3 cornerOne = new Vector3(), Vector3 cornerTwo = new Vector3()){
        Vector3[] linePointPositions;
```

```csharp
            switch(linkType){
                case LinkType.Direct:
                    //直连只需要1根直线，有2个端点
                    // **************
                    line.positionCount = 0;
                    linePointPositions = new Vector3[2];
                    linePointPositions[0] = firstTile.position;
                    linePointPositions[1] = secondTile.position;
                    line.positionCount = linePointPositions.Length;
                    line.SetPositions(linePointPositions);
                    //使用协同程序延迟直线销毁
                    StartCoroutine(DestroyLine());
                break;

                case LinkType.OneCorner:
                    line.positionCount = 0;
                    //二连需要3个端点，中间的点为折点位置
                    //    ********
                    //           *
                    //           *
                    linePointPositions = new Vector3[3];
                    linePointPositions[0] = firstTile.position;
                    linePointPositions[1] = cornerOne;
                    linePointPositions[2] = secondTile.position;
                    line.positionCount = linePointPositions.Length;
                    line.SetPositions(linePointPositions);
                    StartCoroutine(DestroyLine());
                break;

                case LinkType.TwoCorner:
                    line.positionCount = 0;
                    //三连需要4个端点，其中中间的2个端点为折点位置
                    //    ********
                    //           *
                    //           *
                    //    ********
                    linePointPositions = new Vector3[4];
                    linePointPositions[0] = firstTile.position;
                    linePointPositions[1] = cornerOne;
                    linePointPositions[2] = cornerTwo;
                    linePointPositions[3] = secondTile.position;
                    line.positionCount = linePointPositions.Length;
                    line.SetPositions(linePointPositions);
                    Debug.Log(firstTile.name + " " + secondTile.name);
                    StartCoroutine(DestroyLine());
                break;
            }
        }
    IEnumerator DestroyLine(){
            yield return new WaitForSeconds(2f);
            line.positionCount = 0;
        }
}
```

（3）在 Link 脚本中进行枚举。Link 脚本中的代码如下：

```
using System.Collections;
using System.Collections.Generic;
using UnityEngine;

public enum LinkType{Direct,OneCorner,TwoCorner}
public class Link : MonoBehaviour
{
    ……
}
```

（4）返回 Unity 界面，将 LineCreator 脚本拖动到 LinkLine 下，并把 LinkLine 赋给 Line 参数，如图 8-48 所示。

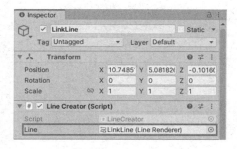

图 8-48　为 Line 参数赋值

（5）返回脚本编辑器，开始绘制直线。Link 脚本中的代码如下：

```
using System.Collections;
using System.Collections.Generic;
using UnityEngine;
LinkType linkType;
public LineCreator line;
public class Link : MonoBehaviour
{
    bool IsLink(int x1,int y1,int x2,int y2){
        //////直连
        //垂直方向
        if(x1 == x2){
            if(Y_LinkDetection(x1,y1,y2)){
                //绘制直线
                linkType = LinkType.Direct;
                line.DrawLinkLine(firSelectedTile.transform,
                                secSelectedTile.transform,linkType);
                return true;
            }
        }
        //水平方向
        if(y1 == y2){
            if(X_LinkDetection(x1,x2,y1)){
                //绘制直线
                linkType = LinkType.Direct;
                line.DrawLinkLine(firSelectedTile.transform,
                                secSelectedTile.transform,linkType);
                return true;
```

```csharp
        }
        ///////二连
        if(OneCornerLinkDetection(x1,y1,x2,y2)){
            //绘制直线
            linkType = LinkType.OneCorner;
            line.DrawLinkLine(firSelectedTile.transform,secSelectedTile.transform,
                              linkType,cornerOnePos);
            return true;
        }
        //////三连
        if(TwoCornerLinkDetection(x1,y1,x2,y2)){
            //绘制直线
            linkType = LinkType.TwoCorner;
            line.DrawLinkLine(firSelectedTile.transform,secSelectedTile.transform,
                              linkType,cornerOnePos,cornerTwoPos);
            return true;
        }
        return false;
    }
    bool OneCornerLinkDetection(int x1, int y1, int x2, int y2){
        //垂直拐角，没有其他图片
        //  (x1,y1)*********
        //         *********
        //  (x1,y2)+*********(x2,y2)
        if(snapshotMap.GetMapItem(y2,x1) == -1){
            if(X_LinkDetection(x1,x2,y2) && Y_LinkDetection(x1,y1,y2)){
                //取得绘制直线的中间拐点
                cornerOnePos = new Vector3(x1+snapshotMap.tileOffset.x,
                                           y2+snapshotMap.tileOffset.y,0);
                return true;
            }
        }
        //水平拐角，没有其他图片
        //  (x1,y1)*********+(x2,y1)
        //         *********
        //         *********(x2,y2)
        if(snapshotMap.GetMapItem(y1,x2) == -1){
            if(X_LinkDetection(x1,x2,y1) && Y_LinkDetection(x2,y1,y2)){
                cornerOnePos = new Vector3(x2+snapshotMap.tileOffset.x,
                                           y1+snapshotMap.tileOffset.y,0);
                return true;
            }
        }
        return false;
    }

    bool TwoCornerLinkDetection(int x1, int y1, int x2, int y2){
        ////右探
        for(int indexX = x1+1; indexX < snapshotMap.GetMapColNum();indexX++){
            //下一个位置没有其他图片
            if(snapshotMap.GetMapItem(y1,indexX) == -1){
                //用当前位置与第二张图片进行二连判断
```

```csharp
            if(OneCornerLinkDetection(indexX,y1,x2,y2)){
                //获得中间两个拐点的位置
                cornerOnePos = new Vector3(indexX+snapshotMap.tileOffset.x,
                                    y1+snapshotMap.tileOffset.y,0);
                cornerTwoPos = new Vector3(indexX+snapshotMap.tileOffset.x,
                                    y2+snapshotMap.tileOffset.y,0);
                return true;
            }
        }
        //如果下一个位置有其他图片，则退出右探遍历
        if(snapshotMap.GetMapItem(y1,indexX) != -1){break;}
    }
    ////左探
    for(int indexX = x1-1;indexX >= 0;indexX--){
        if(snapshotMap.GetMapItem(y1,indexX) == -1){
            if(OneCornerLinkDetection(indexX,y1,x2,y2)){
                cornerOnePos = new Vector3(indexX+snapshotMap.tileOffset.x,
                                    y1+snapshotMap.tileOffset.y,0);
                cornerTwoPos = new Vector3(indexX+snapshotMap.tileOffset.x,
                                    y2+snapshotMap.tileOffset.y,0);
                return true;
            }
        }
        if(snapshotMap.GetMapItem(y1,indexX) != -1){break;}
    }
    ////上探
    for(int indexY = y1+1;indexY < snapshotMap.GetMapRowNum();indexY++){
        if(snapshotMap.GetMapItem(indexY,x1) == -1){
            if(OneCornerLinkDetection(x1,indexY,x2,y2)){
                cornerOnePos = new Vector3(x1+snapshotMap.tileOffset.x,
                                    indexY+snapshotMap.tileOffset.y,0);
                cornerTwoPos = new Vector3(x2+snapshotMap.tileOffset.x,
                                    indexY+snapshotMap.tileOffset.y,0);
                return true;
            }
        }
        if(snapshotMap.GetMapItem(indexY,x1) != -1){break;}
    }
    ////下探
    for(int indexY = y1-1;indexY >= 0;indexY--){
        if(snapshotMap.GetMapItem(indexY,x1) == -1){
            if(OneCornerLinkDetection(x1,indexY,x2,y2)){
                cornerOnePos = new Vector3(x1+snapshotMap.tileOffset.x,
                                    indexY+snapshotMap.tileOffset.y,0);
                cornerTwoPos = new Vector3(x2+snapshotMap.tileOffset.x,
                                    indexY+snapshotMap.tileOffset.y,0);
                return true;
            }
        }
        if(snapshotMap.GetMapItem(indexY,x1) != -1){break;}
    }
    return false;
}
```

（6）返回 Unity 界面，将 LinkLine 赋给 Line 参数，如图 8-49 所示。运行 Unity，发现游戏运行成功。图 8-50～图 8-52 所示分别为直连、二连和三连成功的效果。

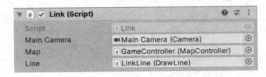

图 8-49　将 LinkLine 赋给 Line 参数

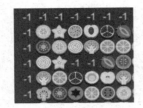 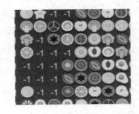 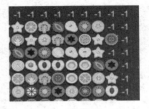

图 8-50　直连成功的效果　　　图 8-51　二连成功的效果　　　图 8-52　三连成功的效果

8.5　本章小结

本章主要介绍了《连连看》游戏的设计与制作，其中重点在于各种连接形式的判断，包括直连、二连和三连等。

8.6　练习题

1．在《连连看》游戏的实现过程中，需要考虑自动判断当前状态是否有解（是否还有可以消除的至少一对图片）的算法：遍历当前状态下所有剩余的图片，为每张图片寻找与之匹配且满足消除条件的另一张图片，若找到则返回真，否则返回假。设当前有 n 张图片，则该算法的时间复杂度为多少？

2．请比较《连连看》游戏打乱图片与《记忆翻牌》游戏中洗牌的算法的区别与联系，并说明各自的优点和缺点。目前《连连看》游戏中打乱图片的实现是否可以改成《记忆翻牌》游戏中洗牌的实现方式？

3．（选做）当前《连连看》游戏的规则是只能进行直连、二连或三连，若允许四连，那么会对游戏产生何种影响？

第 9 章

三消

9.1 游戏简介

2000 年,《宝石迷阵》游戏问世。该游戏的基本玩法如下:玩家需要通过交换相邻的宝石,让同一种宝石形成横向或纵向连续的 3 个或更多个。当形成这样的组合时,宝石便会消失,同时上方会落下新的宝石填充空格,这便是所谓的"瀑布效应"。如果连续形成多个这样的组合,则被称为"连击",可以获得更多积分。《宝石迷阵》是深受大众欢迎的游戏,2002 年便已经成为世界计算机游戏界的"名人",与《俄罗斯方块》游戏齐名。这个系列有 5 部作品,玩家超过 5 亿人,几乎在所有主流平台上线,是《三消》游戏界的翘楚。《宝石迷阵》游戏的画面如图 9-1 所示。

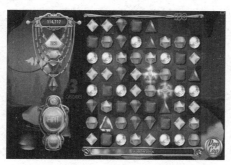

图 9-1 《宝石迷阵》游戏的画面

9.2 游戏规则

玩家选择两个宝石进行位置互换,互换后如果横向或纵向有 3 个或 3 个以上相同的宝石,则消除这几个相同的宝石,如果互换后没有可以消除的宝石,则选中的两个宝石换回原来的位置。宝石消除后的空格由上面掉下来的宝石填充,如图 9-2 所示。每次消除宝石后玩家都能得到一定的分数。

- 连锁:玩家消除宝石后,上面的宝石掉下来填充空格。如果这时游戏池中有连续摆放(横向、纵向)的

图 9-2 《三消》游戏的规则

3个或3个以上相同的宝石，则可以消除这些宝石，这便是一次连锁。空格被新的宝石填充，又可以进行下一次连锁。每次连锁会增加得分。
- **重排**：当玩家已经不能消除任何宝石时，将清空游戏池，并用新的宝石填充。

9.3 游戏开发核心思路

9.3.1 编写 Gemstone 类

可以采用面向对象的思路编写 Gemstone 类。该类包括属性数据和行为。其中，属性数据用于记录宝石在地图数据结构中行和列的位置（rowIndex,colIndex），该宝石的 ID，以及场景中宝石与宝石之间的位置偏移量（xOffset,yOffset）。行为用来更新宝石的行和列，更新宝石在场景中的位置，监听鼠标左键的按下，以及判断选择与否。

9.3.2 初始化游戏场景和地图

游戏场景的生成包括宝石地图的数据结构及宝石的随机生成。
- **宝石地图的数据结构**：可以使用列表（动态数组）存储，数组中的每个元素代表一个宝石，每条列表为一行；一行构建完成后将此条列表保存到头链表（即图 9-3 中的"头列表"）中；可以采用行列索引号对元素进行索引，如通过行号[1]、列号[1]找到的元素对应列表中的[1][1]。

此处使用动态数组（列表）的原因如下：动态数组可以实时改变数组长度，并且可以在指定位置插入新的元素；可以自由控制每行和每列的宝石数量；宝石需要实时且动态地生成、消除和下落。
- **宝石的随机生成**：在宝石种类数组中随机生成一个索引 index，用来随机填充不同的宝石；将填充的宝石放置在场景中对应的位置（列表索引与场景中宝石位置的对应关系如图 9-3 所示）。

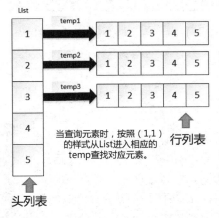

图 9-3 列表索引与场景中宝石位置的对应关系

此时需要注意地图数据结构索引号与宝石在场景中的位置的对应关系，如图 9-4 所示。

图 9-4　地图数据结构索引号与宝石在场景中的位置的对应关系

9.3.3　消除检测

消除检测的流程图如图 9-5 所示。

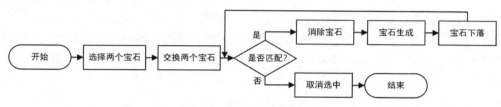

图 9-5　消除检测的流程图

1．选择宝石

在消除时，判断两次选择的宝石是否是相邻的，可以用以下公式确定：

|第一个宝石的行 – 第二个宝石的行| + |第一个宝石的列 – 第二个宝石的列|

如果公式的值为 1，则表示两个宝石相邻。假设第一次选择的宝石的位置为[1,1]。如果选择的第二个宝石的位置为[1,3]，根据公式得到的值为 2，则不能交换；如果选择的第二个宝石的位置为[0,2]，根据公式得到的值为 2，则不能交换，也就是不能对角交换；如果选择的第二个宝石的位置为[0,1]、[1,0]、[2,1]或[1,2]，根据公式得到的值为 1，则可以交换。选择宝石的流程图如图 9-6 所示。

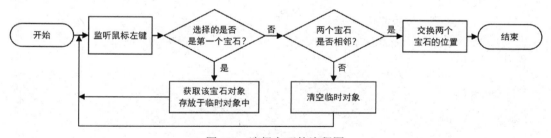

图 9-6　选择宝石的流程图

2．交换宝石的位置

当判断两次选择的宝石相邻时，则将宝石的位置进行交换。首先需要交换两个宝石在列表中的位置，然后交换宝石的行列索引号，接着更新两个宝石在场景中的位置，最后进行三连匹配判断。交换宝石位置的流程图如图 9-7 所示。

3．匹配检测

在进行匹配检测时，首先需要确定匹配中心，具体流程如下。

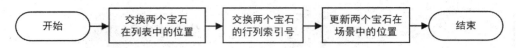

图 9-7 交换宝石位置的流程图

（1）将遍历到的每个宝石确定为匹配中心。

（2）检测匹配中心和右侧两个单位是否同类。

（3）进行横向遍历检测：从左上角开始，横向遍历，如果符合条件，则把这些宝石保存到匹配的列表中。

（4）进行纵向遍历检测：从左上角开始，纵向遍历，如果符合条件，则把这些宝石保存到匹配的列表中。

4．自动匹配与连锁反应

当游戏开始运行时，需要先进行匹配操作，判断生成的场景中是否有三连情况。如果有，则执行消除匹配宝石操作。在运行游戏过程中，在执行一次消除操作后，需要再次判断刷新后的场景中是否仍然有三连情况，如果有，则需要继续执行消除操作，直到场景中没有三连情况。

9.3.4 消除宝石

消除宝石总体上可以分成 3 步：消除匹配的宝石、下移上方的宝石、创建新宝石。

（1）取出匹配列表中的一个宝石。

（2）上方所有宝石向下移动一格。

① 在地图列表中取出待删除宝石上方一个位置的宝石，将其设为 tempGemstone。

② 把 tempGemstone 的行号减 1。

③ 把 tempGemstone 放入修改位置后的地图列表中，覆盖下方（在列表中）的宝石。

④ 更新 tempGemstone 在场景中的位置。

⑤ 回到步骤①，直到最上面的一行。

（3）销毁在匹配列表中取出的宝石。

（4）在顶行创建一个新的宝石。

（5）回到步骤（1），取出下一个匹配的宝石。

伪代码如下：

```
If(是否匹配){
    RemoveMatchedStones();
}
Void RemoveMatchedStones(){
    for(遍历匹配列表){
        RemoveGemstone(matchedStone);
    }
    清空匹配列表
    递归调用，判断其他的相连情况
}

RemoveGemstone(matchedStone){
    1.matchedStone 上方的宝石往下移动一格
    2.销毁 matchedStone
    3.在最顶行创建一个新宝石
}
```

当宝石向下移动一格时，被移动的宝石在场景和地图列表中更新前后的情况如图 9-8 所示。

9.3.5 重新洗牌

如果场景中已经没有可以消除的宝石，则可以将场景中的宝石重新洗牌。

重新洗牌的具体方法如下：先随机对调场景中的宝石，再配对消除。

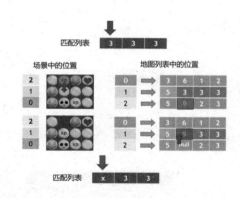

图 9-8 当宝石向下移动一格时，被移动的宝石在场景和地图列表中更新前后的情况

9.3.6 游戏流程图

游戏流程图如图 9-9 所示。

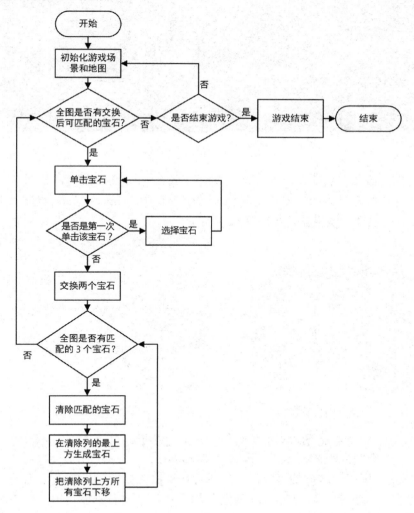

图 9-9 游戏流程图

9.4 游戏实现

9.4.1 导入资源与设置场景

（1）新建工程。新建名为 Eliminate 的工程，选择 2D 选项，如图 9-10 所示。在打开 Eliminate 工程后，按照惯例，在 Assets 文件夹下新建 Sprites 文件夹、Scripts 文件夹和 Prefabs 文件夹，这 3 个文件夹分别用来存储图片、脚本和预制体。

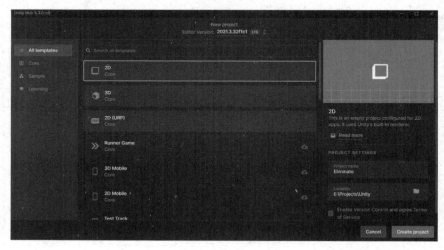

图 9-10　新建工程

（2）导入图片素材。将下载好的图片素材保存到 Sprites 文件夹中，如图 9-11 所示。

（3）对图片进行切割。选中图片，在 Inspector 窗口中将 Sprite Mode 设置为 Multiple，单击 Sprite Editor 按钮打开编辑器，如图 9-12 所示。

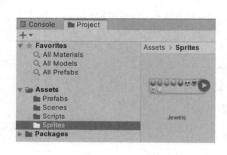

图 9-11　导入图片素材

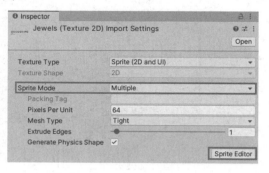

图 9-12　对图片进行切割

（4）如图 9-13 所示，在 Sprite Editor 窗口中，单击 Slice 下拉按钮，将 Type 设置为 Grid By Cell Count。因为图片有 8 列 2 行，所以将 Column 设置为 8，Row 设置为 2，再依次单击 Slice 按钮和 Apply 按钮，这样就可以将宝石图片切割成单独的。切割效果如图 9-14 所示。

（5）设置宝石显示单位。如图 9-15 所示，为了让每个宝石刚好占用一个 Unity 单位，先选中图片，将 Inspector 窗口中的 Pixels Per Unit 设置为 64，再单击 Apply 按钮。

（6）设置摄像机。将摄像机的位置设置为（5.4,5.5），如图 9-16 所示。

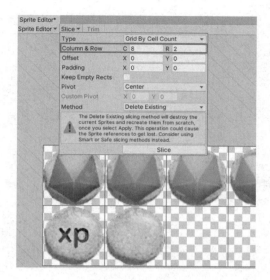

图 9-13　设置切割参数

图 9-14　切割效果

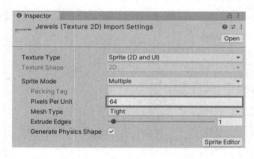

图 9-15　设置宝石显示单位

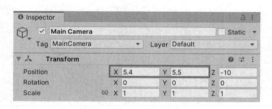

图 9-16　设置摄像机

9.4.2　编写 Gemstone 类

（1）在 Scripts 文件夹下新建 Gemstone 脚本，该脚本用来存储宝石的属性数据和行为。Gemstone 脚本中的代码如下：

```
using System.Collections;
using System.Collections.Generic;
using UnityEngine;

public class Gemstone : MonoBehaviour
{
    public int rowIndex;                            //宝石在地图数据结构中的行号
    public int colIndex;                            //宝石在地图数据结构中的列号
    public float xOffset = 1f, yOffset = 1f;        //两个宝石在场景中的位置间隔
    public int ID;                                  //宝石 ID
    public bool isSelected = false;
```

```
void Start(){

}

//更新宝石的行号和列号，以及宝石在场景中的位置
public void UpdatePositionWithRowCol(int rowIndex, int colIndex){
    //更新行号和列号
    this.rowIndex = rowIndex;
    this.colIndex = colIndex;
    //更新宝石在场景中的坐标
    UpdatePosition();
}

public void UpdatePosition(){
    transform.position = new Vector3(colIndex * xOffset, rowIndex * yOffset, 0);
}

void OnMouseDown(){
    //选择宝石
}
}
```

（2）创建 Gemstone 游戏对象。返回 Unity 界面，创建一个物体用来挂载 Gemstone 脚本。新建一个空游戏对象并将其命名为 Gemstone，如图 9-17 所示。先为 Gemstone 对象添加 Sprite Renderer 组件和 Circle Collider 2D 组件，分别用于挂载宝石图片、添加碰撞盒实现 OnMouseDown() 函数；再为 Gemstone 对象加上 Gemstone 脚本。

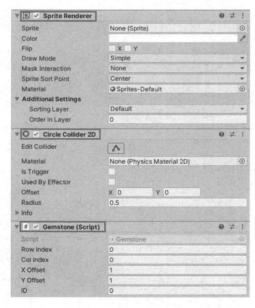

图 9-17　创建 Gemstone 游戏对象

（3）测试 Gemstone 对象的碰撞盒大小是否合适。可以先拖一张宝石的图片赋给 Sprite Renderer 组件的 Sprite 参数，观察场景中的碰撞盒是否完整包裹了宝石。图 9-18 所示为碰撞盒大小合适的情况。

（4）把 Gemstone 对象拖到 Prefabs 文件夹下使其成为一个预制体，如图 9-19 所示，删除场景中的原物体。

图 9-18　碰撞盒大小合适的情况

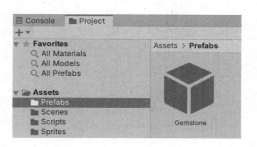

图 9-19　制作 Gemstone 预制体

9.4.3　初始化地图数据结构和场景

（1）在 Scripts 文件夹下新建一个脚本并将其命名为 GameController。GameController 脚本中的代码如下：

```
using System.Collections;
using System.Collections.Generic;
using UnityEngine;

public class GameController : MonoBehaviour
{
    public int rowNum = 7;                        //行数
    public int colNum = 8;                        //列数
    public ArrayList gemstoneList;                //定义行列表
    public Sprite[] gemstoneSprite;               //宝石图片
    public Gemstone gemstonePrefab;               //保存宝石预制体
    // Start is called before the first frame update
    void Start()
    {
        InitMap();                                //初始化地图

    }

    // Update is called once per frame
    void Update()
    {

    }

    void InitMap()
    {
        gemstoneList = new ArrayList();           //实例化列表
        //创建地图数据结构
        for (int row = 0; row < rowNum; row++){
            //创建一行用于存储宝石的列表
            ArrayList newLineList = new ArrayList();
            for (int col = 0; col < colNum; col++){
                //创建一个新的宝石
```

```
                //因为需要对宝石进行更多的设置,所以用一个函数来封装
                Gemstone newGemstone = CreateGemstone(row, col);
                //把新创建的宝石添加到该行的列表中
                newLineList.Add(newGemstone);
            }
            //在生成一行宝石后,把该行的宝石列表添加到头列表中
            gemstoneList.Add(newLineList);
        }
    }

    private Gemstone CreateGemstone(int row, int col){
        Gemstone gemstone = Instantiate(gemstonePrefab) as Gemstone;
        //把生成的宝石挂载到 GameController 游戏对象上,使 Hierarchy 窗口更简洁
        gemstone.transform.SetParent(this.transform);

        //随机生成宝石 ID
        gemstone.ID = Random.Range(0, gemstoneSprite.Length);
        //根据 ID 为宝石对象添加图片
        gemstone.GetComponent<SpriteRenderer>().sprite=gemstoneSprite[gemstone.ID];
        //为了方便管理,为每个宝石命名
        gemstone.name = gemstoneSprite[gemstone.ID].name;
        //设置宝石的位置
        gemstone.UpdatePositionWithRowCol(row, col);
        return gemstone;
    }
}
```

(2)为 GameController 脚本赋值。返回 Unity 界面,新建一个空物体并将其命名为 GameController,把 GameController 脚本赋给该物体。用处理过的宝石图片对 gemstoneSprite 进行赋值。把 Gemstone 预制体赋给 Gemstone Prefab,如图 9-20 所示。

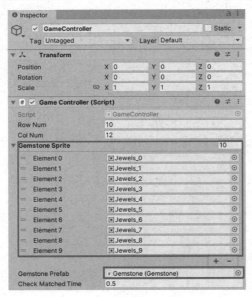

图 9-20 为 Gemstone Prefab 参数赋值

(3)单击"运行"按钮,可以看到已经成功生成游戏初始地图,如图 9-21 所示。

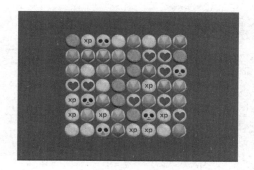

图 9-21　生成的游戏初始地图

9.4.4　选择宝石

（1）实现选择宝石的功能。返回 GameController 脚本，编写 SelectGemstone()函数实现选择宝石的功能。为了在选择时能有明显的反馈，将被选中的宝石修改为红色高亮。GameController 脚本中的代码如下：

```
using System.Collections;
using System.Collections.Generic;
using UnityEngine;

public class GameController : MonoBehaviour
{
    ……

    private Gemstone selectedGemstone;   //临时保存被选中的第一个宝石

    ……

    public void SelectGemstone(Gemstone SelectingGemstone)
    {
        if (selectedGemstone == null)
        {   //选择第一个宝石
            //将当前被选中的宝石赋值到临时变量中
            selectedGemstone = SelectingGemstone;
            //标记当前被选中的宝石的状态
            selectedGemstone.isSelected = true;
            Debug.Log("First Gemstone = " + selectedGemstone.name);
        }
        else
        {   //选择第二个宝石
            if (Mathf.Abs(selectedGemstone.rowIndex-SelectingGemstone.rowIndex)
              +Mathf.Abs(selectedGemstone.colIndex - SelectingGemstone.colIndex)
              == 1)
            {
                Debug.Log("second Gemstone = " + SelectingGemstone.name);
                //交换两个宝石的位置，并进行三连匹配
                Debug.Log("可以交换宝石的位置");
            }
            //清空临时变量
            selectedGemstone.isSelected = false;
```

```
            selectedGemstone = null;
        }
    }
}
```

(2)在 Gemstone 脚本的 OnMouseDown()函数中调用 SelectGemstone()函数，即宝石每被单击一次，就调用一次 SelectGemstone()函数。Gemstone 脚本中的代码如下：

```
using System.Collections;
using System.Collections.Generic;
using UnityEngine;

public class Gemstone : MonoBehaviour
{
    ......

    private SpriteRenderer spriteRenderer;
    public bool isSelected
    {                        //记录该宝石是否已被选中
        set
        {
            //根据传进来的值为 true 或 false 改变宝石状态
            if (value)
            {
                //表示被选中
                spriteRenderer.color = Color.red;
            }
            else
            {
                //表示取消选中
                spriteRenderer.color = Color.white;
            }
        }
    }

    private GameController gameController;

    void Start(){
        gameController = GameObject.Find("GameController").
                     GetComponent<GameController>();
        spriteRenderer = GetComponent<SpriteRenderer>();
    }

    void OnMouseDown(){
        //选择该宝石
        gameController.SelectGemstone(this);
    }
}
```

(3)选择宝石后的效果。返回 Unity 界面进行调试，可以看到，被选中的宝石变成红色。另外，在选中两个宝石后，如果它们相邻，则在 Console 窗口中输出"可以交换宝石的位置"，如图 9-22 所示。

(4)实现交换宝石的位置的功能。打开 GameController 脚本，在实现选择宝石的函数中，判断宝石可以进行交换后就交换它们的位置。GameController 脚本中的代码如下：

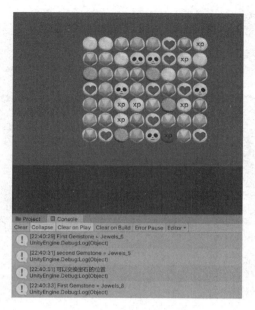

图 9-22 选择宝石后的效果

```
using System.Collections;
using System.Collections.Generic;
using UnityEngine;

public class GameController : MonoBehaviour
{
    ……
    public void SelectGemstone(Gemstone SelectingGemstone)
    {
        if (selectedGemstone == null)
        { //选择第一个宝石
           ……
        }
        else
        { //选择第二个宝石
            if (Mathf.Abs(selectedGemstone.rowIndex - SelectingGemstone.rowIndex) +
              Mathf.Abs(selectedGemstone.colIndex - SelectingGemstone.colIndex)
              == 1)
            {
              Debug.Log("second Gemstone = " + SelectingGemstone.name);
              //交换两个宝石的位置，并进行三连匹配
              Debug.Log("可以交换宝石的位置");
              ExchangeGemstone(selectedGemstone,SelectingGemstone);
            }
            //清空临时变量
            selectedGemstone.isSelected = false;
            selectedGemstone = null;
        }
    }

    //交换宝石的位置
    void ExchangeGemstone(Gemstone stone1, Gemstone stone2)
```

```csharp
{
    //交换宝石在地图列表中的位置
    //把宝石 2 放到列表中原来放置宝石 1 的位置
    AddGemstoneToMap(stone1.rowIndex, stone1.colIndex, stone2);
    //把宝石 1 放到列表中原来放置宝石 2 的位置
    AddGemstoneToMap(stone2.rowIndex, stone2.colIndex, stone1);
    //交换宝石行号
    int tempRowIndex = stone1.rowIndex;
    stone1.rowIndex = stone2.rowIndex;
    stone2.rowIndex = tempRowIndex;
    //交换宝石列号
    int tempColIndex = stone1.colIndex;
    stone1.colIndex = stone2.colIndex;
    stone2.colIndex = tempColIndex;
    //更新宝石在场景中的位置
    stone1.UpdatePosition();
    stone2.UpdatePosition();
}

void AddGemstoneToMap(int rowIndex, int colIndex, Gemstone stone)
{
    //取出对应的行所在的列表
    ArrayList temp = gemstoneList[rowIndex] as ArrayList;
    //把宝石放到对应的列表位置上
    temp[colIndex] = stone;
}
```

（5）返回 Unity 界面进行测试。可以看到，单击两个相邻宝石可以实现交换，交换宝石的效果如图 9-23 所示。

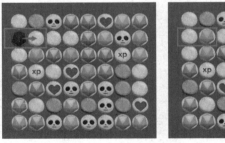
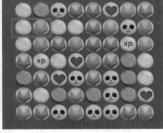

图 9-23　交换宝石的效果

9.4.5　消除宝石

（1）实现匹配判断的功能。在 GameController 脚本中实现用于匹配判断的函数 Match()。GameController 脚本中的代码如下：

```csharp
using System.Collections;
using System.Collections.Generic;
using UnityEngine;

public class GameController : MonoBehaviour
{
```

```csharp
......
    private ArrayList matchedList;                    //临时保存需要被消除的宝石
    private Gemstone stone1, stone2, stone3;          //临时保存需要匹配的宝石

    public void SelectGemstone(Gemstone SelectingGemstone)
    {
        if (selectedGemstone == null)
        {   //选择第一个宝石
            ......
        }
        else
        {   //选择第二个宝石
            if (Mathf.Abs(selectedGemstone.rowIndex - SelectingGemstone.rowIndex) +
                Mathf.Abs(selectedGemstone.colIndex - SelectingGemstone.colIndex)
                == 1)
            {
                //Debug.Log("second Gemstone = " + SelectingGemstone.name);
                //Debug.Log("可以交换宝石的位置");
                //ExchangeGemstone(selectedGemstone,SelectingGemstone);
                //交换两个宝石的位置，并进行三连匹配
                ExchangeGemstoneAndMatch(selectedGemstone, SelectingGemstone);
            }
            //清空临时变量
            selectedGemstone.isSelected = false;
            selectedGemstone = null;
        }
    }

    bool Match()
    {
        //进行横向检测和纵向检测
        if (CheckHorizontalMatch() || CheckVerticalMatch())
        {
            return true;
        }
        return false;
    }

    //横向检测
    bool CheckHorizontalMatch()
    {
        for (int row = 0; row < rowNum; row++)
        {
            //因为在检测时需要与后面的两个位置进行匹配，所以这里要求colIndex < colNum-2
            for (int col = 0; col < colNum - 2; col++)
            {
                //检测中心和右侧两个单位是否为同类
                stone1 = GetGemstoneFromMap(row, col);
                stone2 = GetGemstoneFromMap(row, col + 1);
                stone3 = GetGemstoneFromMap(row, col + 2);
                if ((stone1.ID == stone2.ID) && (stone1.ID == stone3.ID))
                {
```

```csharp
                AddToMatchedList(stone1);
                AddToMatchedList(stone2);
                AddToMatchedList(stone3);
                return true;
            }
        }//col
    }//row
    return false;
}

//纵向检测
bool CheckVerticalMatch()
{
    for (int col = 0; col < colNum; col++)
    {
        for (int row = 0; row < rowNum - 2; row++)
        {
            stone1 = GetGemstoneFromMap(row, col);
            stone2 = GetGemstoneFromMap(row + 1, col);
            stone3 = GetGemstoneFromMap(row + 2, col);
            if (stone1.ID == stone2.ID && stone1.ID == stone3.ID)
            {
                AddToMatchedList(stone1);
                AddToMatchedList(stone2);
                AddToMatchedList(stone3);
                return true;
            }
        }//row
    }//col
    return false;
}

//从地图中获取宝石
Gemstone GetGemstoneFromMap(int rowIndex, int colIndex)
{
    ArrayList temp = gemstoneList[rowIndex] as ArrayList;
    return temp[colIndex] as Gemstone;
}

//添加到匹配列表中
void AddToMatchedList(Gemstone stone)
{
    if (matchedList == null)
    {
        matchedList = new ArrayList();
    }
    //检测宝石是否已经在列表中
    //如果没有则返回-1，可以把当前的宝石加入列表中
    int index = matchedList.IndexOf(stone);
    if (index == -1)
    {
        matchedList.Add(stone);
    }
```

```
    }
    void ExchangeGemstoneAndMatch(Gemstone stone1, Gemstone stone2)
    {
        ExchangeGemstone(stone1, stone2);
        if (Match())
        {
            Debug.Log("Remove Matched");
        }
        else
        {
            //不匹配时再调换回去
            ExchangeGemstone(stone1, stone2);
        }
    }
}
```

（2）返回 Unity 界面进行调试。可以发现，当横向检测和纵向检测都不满足时，即使选中相邻的宝石也不会进行交换。

（3）实现删除匹配宝石、下移上方宝石和创建新宝石的功能。GameController 脚本中的代码如下：

```
using System.Collections;
using System.Collections.Generic;
using UnityEngine;

public class GameController : MonoBehaviour
{
......
void ExchangeGemstoneAndMatch(Gemstone stone1, Gemstone stone2)
    {
        ExchangeGemstone(stone1, stone2);
        if (Match())
        {
            Debug.Log("Remove Matched");
            RemoveMatchedStones();
        }
        else
        {
            //不匹配时再调换回去
            ExchangeGemstone(stone1, stone2);
        }
    }

    //删除匹配的宝石
    void RemoveMatchedStones()
    {
        //1.取出匹配列表中的宝石
        //遍历整个匹配列表
        for (int index = 0; index < matchedList.Count; index++)
        {
            //删除宝石
            RemoveGemstone(matchedList[index] as Gemstone);
        }
```

```
        matchedList.Clear();
    }

    void RemoveGemstone(Gemstone matchedGemstone)
    {
        //Debug.Log("RemoveGemstone");
        //2.上方所有宝石下移
        //从当前被消除宝石的位置的上一个位置开始遍历
        for (int row = matchedGemstone.rowIndex + 1; row < rowNum; row++)
        {
            //获得被消除宝石上方的一个宝石
            Gemstone tempGemstone = GetGemstoneFromMap(row, matchedGemstone.colIndex);
            //往下移动一行
            tempGemstone.rowIndex--;
            //把要下移的宝石放到地图中的下一个位置
            AddGemstoneToMap(tempGemstone.rowIndex, tempGemstone.colIndex, tempGemstone);
            tempGemstone.UpdatePositionWithRowCol(tempGemstone.rowIndex,
                                                  tempGemstone.colIndex);
        }
        //3.销毁当前宝石
        Destroy(matchedGemstone.gameObject);
        //4.在顶行创建一个新的宝石
        Gemstone newGemstone=CreateGemstone(rowNum,
                                             matchedGemstone.colIndex);
        newGemstone.name = newGemstone.name + "New";
        newGemstone.rowIndex--;
        AddGemstoneToMap(newGemstone.rowIndex, newGemstone.colIndex,
                         newGemstone);
        newGemstone.UpdatePosition();
    }
}
```

（4）返回 Unity 界面进行调试。在消除匹配宝石后，上方宝石会移到下方，顶行会出现新的宝石。图 9-24 所示为当前游戏运行效果。

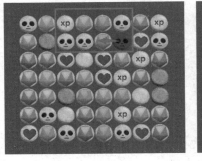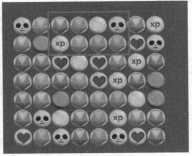

图 9-24 当前游戏运行效果

（5）游戏初始匹配判断及连锁反应。在地图刚生成时就需要进行匹配判断，以消除满足三连条件的宝石。在消除满足三连条件的宝石后，上方宝石的下落及顶行新创建的宝石可能会构成新的三连，这时需要连锁地进行匹配判断和消除。GameController 脚本中的代码如下：

```
using System.Collections;
using System.Collections.Generic;
```

```csharp
using UnityEngine;

public class GameController : MonoBehaviour
{
    ……

    public float CheckMatchedTime = 0.5f;

    void Start()
    {
        InitMap();   //初始化地图

        //游戏刚开始时进行匹配
        if (Match())
        {
            RemoveMatchedStones();
        }
    }

    ……

    //删除匹配的宝石
    void RemoveMatchedStones()
    {
        //取出匹配列表中的宝石
        //遍历整个匹配列表
        for (int index = 0; index < matchedList.Count; index++)
        {
            //删除宝石
            RemoveGemstone(matchedList[index] as Gemstone);
        }
        matchedList.Clear();
        StartCoroutine(WaitForCheckMatchedAgain());
    }

    IEnumerator WaitForCheckMatchedAgain()
    {
        yield return new WaitForSeconds(CheckMatchedTime);
        if (Match())
        {
            RemoveMatchedStones();
        }
    }
}
```

（6）返回 Unity 界面进行调试。此时如果没有出错，那么在地图生成时，满足三连条件的宝石自动被消除。在交换宝石后，宝石的消除和再匹配也是连锁的。

9.4.6　重新洗牌操作

当游戏进行到一定程度后，场景中可能已经没有可以消除的宝石，这时需要重新对场景中的宝石进行洗牌。

（1）在 GameController 脚本中编写洗牌的函数。按下 R 键，调用洗牌函数进行洗牌。GameController 脚本中的代码如下：

```csharp
using System.Collections;
using System.Collections.Generic;
using UnityEngine;

public class GameController : MonoBehaviour
{
    ......

    void Update()
    {
        if(Input.GetKeyDown(KeyCode.R)){
            RefreshMap();
        }
    }
    ......

    public void RefreshMap()
    {
        for (int row = 0; row < rowNum; row++)
        {
            for (int col = 0; col < colNum; col++)
            {
                Gemstone stone1 = GetGemstoneFromMap(row, col);
                Gemstone stone2 = GetGemstoneFromMap(Random.Range(0,
                                    rowNum),Random.Range(0, colNum));
                AddGemstoneToMap(stone1.rowIndex, stone1.colIndex, stone2);
                AddGemstoneToMap(stone2.rowIndex, stone2.colIndex, stone1);
                int tempRowIndex = stone1.rowIndex;
                stone1.rowIndex = stone2.rowIndex;
                stone2.rowIndex = tempRowIndex;

                int tempColIndex = stone1.colIndex;
                stone1.colIndex = stone2.colIndex;
                stone2.colIndex = tempColIndex;
                Debug.Log("Exchange");
                stone1.UpdatePosition();
                stone2.UpdatePosition();
                Debug.Log("RefreshMap Done");
            }//col
        }//row
        if (Match())
        {
            RemoveMatchedStones();
        }
    }
}
```

（2）返回 Unity 界面，当按下 R 键时，可以看到地图被重新洗牌，重新洗牌的效果如图 9-25 所示。

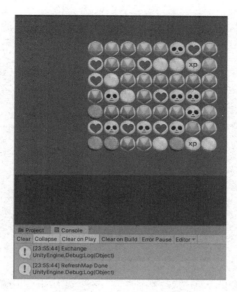

图 9-25 重新洗牌的效果

9.5 本章小结

本章介绍了宝石消除类游戏的设计与实现,其中重点在于消除条件的检测,以及宝石消除后的连锁反应如何实现。

9.6 练习题

1. 在本章介绍的宝石消除类游戏中,宝石阵列的数据结构如下:以行为单位保存到列表中,并将所有行保存到总列表中。现考虑另一种方式,即以列为单位保存到列表中,并将所有列保存到总列表中,若有 n 列,则在执行删除宝石时(暂不考虑更新宝石的位置信息)的时间复杂度为多少?

2. 在本章介绍的宝石消除类游戏中,宝石的相邻性判断以四邻域(上、下、左、右)为基准,此时采用的相邻判断条件为曼哈顿街区距离是否为 1。若改成八邻域(在四邻域的基础上加入左上、左下、右上、右下),则判断相邻的条件应改为什么?

3. (选做)在本章介绍的宝石消除类游戏中,地图为规则排列的阵列形式,可以用列表来表示。若地图中出现"洞"(在阵列中某些点上为障碍物,而非宝石),则应该如何修改?请简要描述思路。

第 10 章

扫雷

10.1 游戏简介

《扫雷》是一款十分经典的大众类益智游戏。该游戏基于简单的数字逻辑来排查画面内所有的地雷。其最原始版本可以追溯到 1973 年推出的《方块》游戏。在推出《方块》游戏不久后，该游戏就被改写成 Rlogic 游戏。在 Rlogic 游戏中，玩家的任务是作为美国海军陆战队队员，为指挥中心探出一条没有地雷的安全路线，如果路全部被地雷堵死便算输。之后，汤姆·安德森在 Rlogic 游戏的基础上又编写出了《地雷》游戏，由此奠定了现代《扫雷》游戏的雏形。1981 年，微软公司的罗伯特·杜尔和卡特·约翰逊在 Windows 3.1 系统上装载了《扫雷》游戏，由此《扫雷》游戏才正式在全世界推广。Windows 版本的《扫雷》游戏如图 10-1 所示。

图 10-1 Windows 版本的《扫雷》游戏

10.2 游戏规则

10.2.1 《扫雷》游戏的布局

《扫雷》游戏的游戏区包括雷区、地雷计数器（记录剩余地雷数）和计时器（记录游戏时间）。在确定大小的雷区中随机布置一定数量的地雷（初级为 9×9 个方格，10 个地雷；中级为 16×16 个方格，40 个地雷；高级为 16×30 个方格，99 个地雷；自定义级别可以自己设定雷区大小和地雷数，但是雷区大小不能超过 24×30 个方格），玩家需要尽快找出雷区中所有不是地雷的方格，且不允许踩到地雷。

10.2.2 《扫雷》游戏的基本操作

《扫雷》游戏的基本操作包括左键单击、右键单击、左右双击 3 种。其中，左键单击用于打开安全的方格，推进游戏进度；右键单击用于标记地雷，以辅助判断，或者为接下来的左右双击做准备；当一个数字周围的地雷标记完时，左右双击相当于对数字周围未打开的方格均执行一次左键单击操作。

- **左键单击**：在判断出不是地雷的方格上按下鼠标左键，可以打开该方格。如果方格上出现数字，则该数字表示以该方格为中心的 3×3 区域中的地雷数；如果方格上为空（相当于 0），则可以递归地打开与空方格相邻的方格；如果不幸触雷，则游戏结束。在游戏区中间/边缘/角落单击方格时数字显示地雷数的范围，小圆点所在处为鼠标左键单击方格。范围示意图如图 10-2 所示。
- **右键单击**：在判断为地雷的方格上按下鼠标右键，可以标记地雷（显示为小红旗）。重复一次或两次操作可取消标记（如果在游戏菜单中选择了"标记"命令，则需要操作两次来取消标记）。
- **左右双击**：同时按下鼠标左键和鼠标右键可以完成左右双击。当双击位置周围已标记地雷数等于该位置数字时操作有效，相当于对该数字周围未打开的方格均进行一次左键单击操作。当地雷未标记完全时使用左右双击无效。若数字周围有标错的地雷，则游戏结束，且标错的地雷上会显示符号"×"。

游戏成功和失败的判定条件如下：当玩家猜出所有地雷时，游戏成功；当一个地雷被踩中时，所有地雷都将显示，游戏失败，画面如图 10-3 所示。

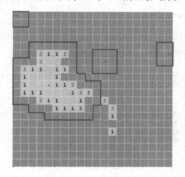

图 10-2　范围示意图

图 10-3　游戏失败画面

10.3　游戏开发核心思路

10.3.1　初始化游戏地图

通过观察整个游戏场景，会自然而然地联想到使用二维数组存储全局地图，并生成给予 Tile 的场景。此步骤还涉及随机生成地雷布局及计算邻近的地雷数。

10.3.2　编写 Tile 类

需要为代表《扫雷》游戏中每个方格的 Tile 编写 Tile 类。在 Tile 类中，需要编写 Tile 的状态、属性和操作。
- **Tile** 的状态：Unreveal（未翻开）、Reveal（翻开）和 Flag（标旗）。
- **Tile** 的属性：Tile 在场景中的位置（int 类型）、是否有地雷，以及是否是鼠标指针悬浮所在对象。
- **Tile** 的操作：设置位置信息、计算邻近方格标识的地雷数、翻开方格、标旗、游戏结束、游戏胜利。

10.3.3 随机生成地雷

在地图初始化时，需要随机生成地雷，具体步骤如图 10-4 所示。

图 10-4 随机生成地雷的步骤

（1）在地图中随机寻找一个没有被设置成有地雷的 Tile。
（2）把该 Tile 设置成有地雷的状态（bool hasMine）。
（3）以该 Tile 为中心，向左、向右、向上、向下获得邻近的 Tile。
（4）把这些邻近 Tile 中记录附近地雷数的变量+1（需要事先初始化为 0）。
（5）把设置成地雷的 Tile 加入地雷列表，以供后续使用。
（6）如果未达到设置的地雷总数，则回到步骤（1）。

获得邻近 Tile 的示意图如图 10-5 所示。如果没有在边界，就需要获得 8 个邻近 Tile；如果在左边界、右边界、上边界或下边界（没有在对角位置），就需要获得 5 个邻近 Tile；如果在对角位置，就需要获得 3 个邻近 Tile。

图 10-5 获得邻近 Tile 的示意图

10.3.4 编写辅助函数，显示地图信息

在编写游戏的时候，为了调试方便、显示直观，可以编写一个地图快照，显示地雷的分布和各个 Tile 附近的地雷数，如图 10-6 所示。

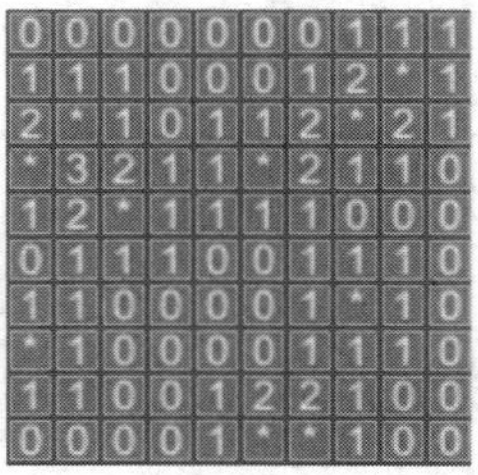

图 10-6 地图快照

10.3.5 Tile 的交互

需要为地图上的各个 Tile 添加各种状态，如未翻开、翻开、标旗、地雷。还需要的交互

是左键单击方格为翻开方格、右键单击方格为标旗或取消标旗。

10.3.6 方格的单击（左键、遍历翻开、右键）

左键单击方格时交互的判断流程如图 10-7 所示，重点是翻开方格时需要遍历且翻开邻近空白方格。

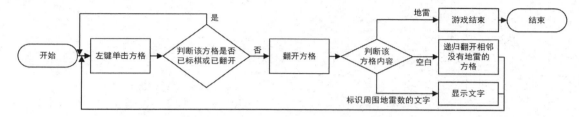

图 10-7 左键单击方格时交互的判断流程

在遍历翻开空白方格时，交互的判断流程如图 10-8 所示。遍历对象是空白方格，即它的四周没有地雷；遍历操作是翻开相邻的方格；遍历停止条件是遍历到的空白方格附近有地雷。当用鼠标右键单击方格时，将当前方格的状态更改为标旗/未翻开。当标识为红旗时，禁用鼠标左键。

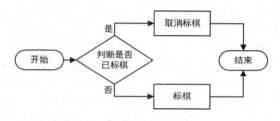

图 10-8 交互的判断流程

10.3.7 游戏结束判断

（1）如果用鼠标左键单击方格显示的是地雷，则游戏失败，并显示所有地雷。

（2）如果所有 Tile 都被翻开，则游戏胜利。可以使用未翻开 Tile 的数量来判断游戏是否胜利，如果未翻开 Tile 的数量为 0，则表示已经翻开了所有 Tile 并且没有翻到地雷，游戏胜利。

10.3.8 UI 控制

UI 效果如图 10-9 所示。UI 包括地雷数显示、计时功能和头像显示。

（1）地雷数显示：若标记一面旗子，则地雷数-1。

（2）计时功能：当游戏开始后玩家第一次单击 Tile 时开始计时，游戏胜利或失败时停止计时。

（3）头像显示如图 10-10 所示。头像共有 4 种：成功翻开 Tile，显示 Ohh 表情；踩到地雷，显示 Dead 表情；标旗，显示 OK 表情；游戏胜利，显示 Wow 表情。

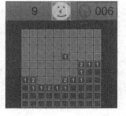

图 10-9 UI 效果

图 10-10 头像显示

10.3.9 游戏流程图

游戏流程图如图 10-11 所示。

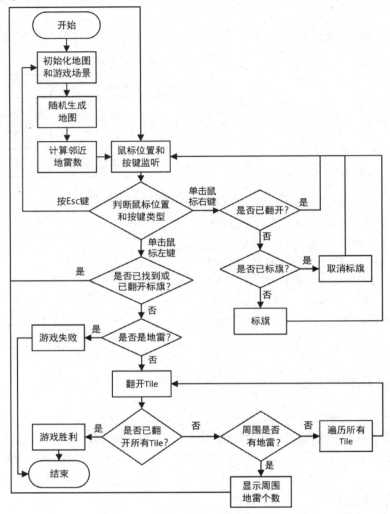

图 10-11 游戏流程图

10.4 游戏实现

10.4.1 资源导入与场景设置

（1）新建工程。新建名为 MineSweeper 的工程，选择 2D 选项，如图 10-12 所示。打开 MineSweeper 工程后，按照惯例，在 Assets 文件夹下新建 Sprites 文件夹、Scripts 文件夹和 Prefabs 文件夹，这 3 个文件夹分别用来保存图片、脚本和预制体。

（2）导入图片素材。将下载的图片素材保存到 Sprites 文件夹中，如图 10-13 所示。

（3）设置单位。为了让每个 Tile 刚好占用一个 Unity 单位，选中所有图片，将 Inspector

窗口中的 Pixels Per Unit 设置为 32（单个 Tile 图片的边长），如图 10-14 所示，单击 Apply 按钮。

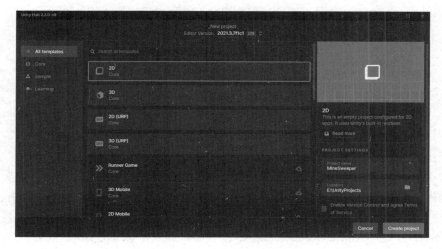

图 10-12　新建工程

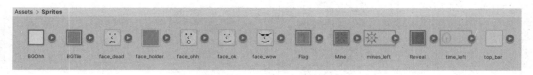

图 10-13　导入图片素材

（4）设置摄像机。将摄像机的位置设置为（4.5，6），Size 设置为 7，如图 10-15 所示。

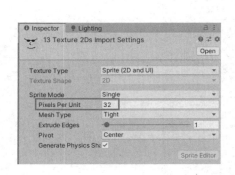

图 10-14　设置单位

图 10-15　设置摄像机

10.4.2　初始化场景

（1）编写 Tile 类和生成地图。

① 编写 Tile 类。在 Scripts 文件夹下新建一个脚本并将其命名为 Tile，该脚本用来存储每个 Tile 的状态、属性和操作。在这个步骤中，先编写涉及 Tile 的位置的相关代码。Tile 脚本中的代码如下：

```
using System.Collections;
```

```csharp
using System.Collections.Generic;
using UnityEngine;

public class Tile : MonoBehaviour
{
    //场景中的位置
    public int posX, posY;

    //设置位置
    public void SetPosition(int x, int y)
    {
        this.posX = x;
        this.posY = y;
    }
}
```

② 制作 Tile 预制体。返回 Unity 界面，创建一个空物体并将其命名为 TilePrefab。为新创建的物体加上 Sprite Renderer 组件和 Box Collider 2D 组件，并附加 Tile 脚本。将 Sprites 文件夹下的 BGTile 赋给 Sprite Renderer 组件的 Sprite 参数。为了使碰撞盒的大小完全包裹图片，将 Box Collider 2D 组件的 Size 参数设置为（1,1），如图 10-16 所示。

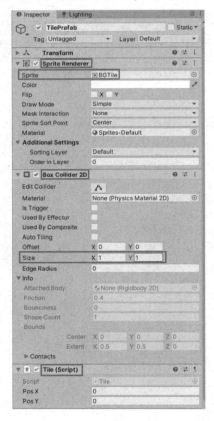

图 10-16　制作 Tile 预制体

③ 将 TilePrefab 物体拖到 Prefabs 文件夹下使其成为一个预制体，并且删除场景中的原物体。

④ 初始化地图。在 Scripts 文件夹下新建一个脚本并将其命名为 Map。Map 脚本中的代码如下：

```csharp
using System.Collections;
using System.Collections.Generic;
using UnityEngine;

public class Map : MonoBehaviour
{
    //地图的行数和列数
    public int rowNum = 10;
    public int colNum = 10;

    public Tile[,] map;                 //地图数组
    public Tile tilePrefab;             //Tile 预制体

    void Start()
    {
        CreateMap();                    //生成地图
    }

    //生成地图
    void CreateMap()
    {
        //1.初始化地图
        CreateTiles();
    }

    //初始化地图
    void CreateTiles()
    {
        map = new Tile[rowNum, colNum];
        for (int row = 0; row < rowNum; row++)
        {
            for (int col = 0; col < colNum; col++)
            {
                //注意置换行号和列号
                Tile newTile = Instantiate(tilePrefab, new Vector3(col, row, 0),
                            Quaternion.identity);
                map[row, col] = newTile;
                map[row, col].SetPosition(col, row);
                //把 Tile 挂载到附加了 Map 脚本的物体上
                map[row, col].transform.SetParent(transform);
            }//col
        }//row
    }
}
```

⑤ 返回 Unity 界面。新建一个空物体并将其命名为 Map，为该物体附加 Map 脚本。把 TilePrefab 物体拖给 Map 脚本的 Tile Prefab 参数，如图 10-17 所示。

⑥ 单击"运行"按钮，生成的地图如图 10-18 所示。

（2）随机生成地雷。

① 在初始化地图后，接下来需要随机生成地雷。先打开 Tile 脚本，编写与初始化地雷有关的参数和函数。Tile 脚本中的代码如下：

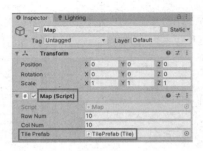

图 10-17　挂载 Map 脚本

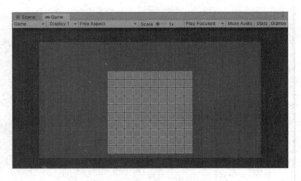

图 10-18　生成的地图

```
using System.Collections;
using System.Collections.Generic;
using UnityEngine;

public class Tile : MonoBehaviour
{
    ……

    //设置该 Tile 是否有地雷
    public bool hasMine = false;
    //该方格周围的方格列表
    private List<Tile> adjacentTiles = new List<Tile>();
    //该方格周围的地雷个数
    private int MineCountNearby = 0;

    private Map map;

    private void Awake()
    {
        if (map == null)
        {
            map = GameObject.Find("Map").GetComponent<Map>();
        }
    }

    //设置为有地雷状态
    public void AddMine()
    {
        //表示该 Tile 上有地雷
        hasMine = true;
        //把该 Tile 对象作为遍历邻近 Tile 的中心点取得邻近的所有方格
        adjacentTiles = map.GetAdjacentTiles(this);
        //遍历有地雷的邻近 Tile，将每个 Tile 的邻近地雷数+1
        foreach (Tile tile in adjacentTiles)
        {
            tile.MineCountNearby++;
        }
    }
}
```

② 打开 Map 脚本，编写用于随机生成地雷的函数 GenerateMines()。Map 脚本中的代码如下：

```csharp
using System.Collections;
using System.Collections.Generic;
using UnityEngine;

public class Map : MonoBehaviour
{
    ……

    public int mineCount = 10;                                  //地雷总数
    private List<Tile> tileWithMines = new List<Tile>();        //用于存储有地雷的 Tile

    void Start()
    {
        CreateMap();                                            //生成地图
    }

    //生成地图
    void CreateMap()
    {
        //1.初始化地图
        CreateTiles();
        //2.初始化地雷
        GenerateMines();
    }

    //初始化地雷
    void GenerateMines()
    {
        //使用 for 循环逐个生成地雷
        for (int mineIndex = 0; mineIndex < mineCount; mineIndex++)
        {
            Tile tileWithMine;
            //随机遍历地图中的 Tile，如果被遍历到的 Tile 已经有地雷，则重新遍历
            do
            {
                tileWithMine = map[Random.Range(0, rowNum),
                                   Random.Range(0, colNum)];
            } while (tileWithMine.hasMine);
            //将遍历到的满足条件（没有地雷）的 Tile 设置为有地雷状态
            tileWithMine.AddMine();
            //将遍历到的有地雷的 Tile 放到地雷列表中
            tileWithMines.Add(tileWithMine);
        }
    }

    //取得以某个方格为中心的周围所有方格对象列表
    public List<Tile> GetAdjacentTiles(Tile centerTile)
    {
        //对周围 8 个方格进行遍历
        //如果被单击的方格处于边缘，则只需要遍历 5 个方格
        //如果被单击的方格处于角落，则只需要遍历 3 个方格
        int startX = centerTile.posX;
        int startY = centerTile.posY;
```

```csharp
        //新建一个列表存储该方格周围所有的方格对象
        List<Tile> adjacentTiles = new List<Tile>();
        //左、右方格的坐标
        for (int dx = -1; dx <= 1; dx++)
        {
            int newX = startX + dx;
            //判断左、右方格是否处于边界
            if (newX < 0 || newX >= rowNum)
            {
                continue;
            }
            //上、下方格的坐标，注意该循环在dx循环中
            for (int dy = -1; dy <= 1; dy++)
            {
                int newY = startY + dy;
                //判断上、下方格是否处于边界
                if (newY < 0 || newY >= colNum)
                {
                    continue;
                }
                adjacentTiles.Add(map[newY, newX]);
            }//end dy
        }//end dx
        return adjacentTiles;
    }
}
```

（3）编写辅助函数，显示地图信息。

① 为了在初始化地图和地雷时确保整个地图的信息设置正确，需要实现一个显示地图信息的辅助函数。在 Tile 脚本中实现一个返回当前 Tile 附近的地雷数的函数 GetNearbyMineCount()。Tile 脚本中的代码如下：

```csharp
using System.Collections;
using System.Collections.Generic;
using UnityEngine;

public class Tile : MonoBehaviour
{
    ……

    //返回当前Tile附近的地雷数
    public int GetNearbyMineCount()
    {
        return MineCountNearby;
    }
}
```

② 在 Map 脚本中实现用于显示辅助信息的函数 ShowNearbyMineCount()。Map 脚本中的代码如下：

```csharp
using System.Collections;
using System.Collections.Generic;
using UnityEngine;
using TMPro;

public class Map : MonoBehaviour
```

```csharp
{
    ......
    //用于打印辅助信息的 TextMesh
    private TextMesh[,] textMeshes;
    public TextMesh textMeshPrefab;
    //用于打印辅助信息的 TextMeshPro
    private TextMeshPro[,] textMeshes;
    public TextMeshPro textMeshPrefab;
    public Transform textMeshRoot;

    //生成地图
    void CreateMap()
    {
        //1.初始化地图
        CreateTiles();
        //2.初始化地雷
        GenerateMines();
        //3.显示辅助信息
        ShowNearbyMineCount();
    }
    //显示辅助信息
    void ShowNearbyMineCount()
    {
        //分配空间
        if (textMeshes == null)
        {
            textMeshes = new TextMesh[rowNum, colNum];
        }
        for (int row = 0; row < rowNum; row++)
        {
            for (int col = 0; col < colNum; col++)
            {
                if (textMeshes[row, col] == null)
                {
                    textMeshes[row, col] = Instantiate(textMeshPrefab,
                            new Vector3(col, row, -1),Quaternion.identity);
                    textMeshes[row, col].transform.SetParent(textMeshRoot);
                    //如果该 Tile 不是地雷,则显示它附近的地雷数
                    if (map[row, col].hasMine == false)
                    {
                        textMeshes[row, col].text = map[row, col].
                                                GetNearbyMineCount().ToString();
                    }
                    else
                    {
                        //如果该 Tile 是地雷,则用 "*" 表示
                        textMeshes[row, col].text = "*";
                    }
                }
            }//col
        }//row
    }
}
```

③ 制作用于打印地图信息的 Text 预制体。返回 Unity 界面，在主菜单栏中选择 GameObject→3D Object→Text→TextMeshPro 命令，创建一个 3D Text 物体并将其命名为 MineCountText。调整 MineCountText 物体的参数，如图 10-19 所示。把 MineCountText 物体拖到 Prefabs 文件夹下，使其成为一个预制体，并删除场景中的原物体。

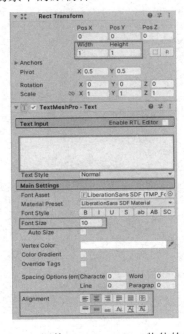

图 10-19　调整 MineCountText 物体的参数

④ 新建一个空物体并将其命名为 MineCountTexts。用上一个步骤制作的 MineCountText 预制体和 MineCountTexts 物体对 Map 脚本进行赋值，如图 10-20 所示。

⑤ 单击"运行"按钮，此时地图信息已经被打印出来，如图 10-21 所示。

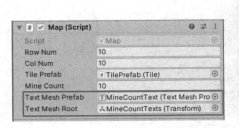

图 10-20　对 Map 脚本进行赋值

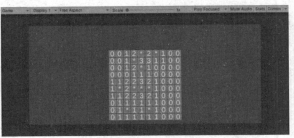

图 10-21　打印的地图信息

10.4.3　左、右键操作

（1）左键翻开 Tile。

① 在 Unity 中修改 TilePrefab 物体，这样在翻开某个 Tile 后，该 Tile 上能显示邻近地雷的信息。双击 TilePrefab 物体，进入预制体修改界面。为 TilePrefab 物体添加一个 3D Text 子物体，该 3D Text 子物体的参数如图 10-22 所示。在设置完成后，将该物体重命名为 MineCountText（需要先关闭该物体，让其不显示出来）。

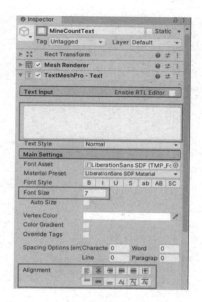

图 10-22　3D Text 子物体的参数

② 在 Tile 脚本中设置 Tile 的 4 种状态。Tile 脚本中的代码如下：

```
using System.Collections;
using System.Collections.Generic;
using UnityEngine;

public class Tile : MonoBehaviour
{
    ……

    //Tile 的 4 种状态：未翻开、翻开、标旗、地雷
    public enum State { Unreveal, Reveal, Flag, Mine };
    //初始化初始状态为未翻开
    State state = State.Unreveal;
    //对应 4 种状态的图片
    public Sprite unrevealSprite, revealSprite, flagSprite, mineSprite;
    private SpriteRenderer spriteRenderer;

    private void Awake()
    {
        if (map == null)
        {
            map = GameObject.Find("Map").GetComponent<Map>();
        }
        spriteRenderer = GetComponent<SpriteRenderer>();
    }

}
```

③ 在 Tile 脚本中通过 OnMouseOver() 函数实现监听按键。Tile 脚本中的代码如下：

```
using System.Collections;
using System.Collections.Generic;
using UnityEngine;
using TMPro;
```

```csharp
public class Tile : MonoBehaviour
{
    ......
    public TextMesh mineCountText;
    public TextMeshPro mineCountText;

    ......

    //鼠标指针在该 Tile 上
    void OnMouseOver()
    {
        //Debug.Log("Over " + posX + " " + posY);
        //左键按下后弹起
        if (Input.GetMouseButtonUp(0))
        {
            Reveal();         //翻开方格
        }
    }

    //翻开方格
    private void Reveal()
    {
        //如果被单击的 Tile 当前状态为翻开，则退出函数
        if (state == State.Reveal) return;
        //如果该 Tile 没有地雷，则可以执行以下操作
        if (hasMine == false)
        {
            state = State.Reveal;
            spriteRenderer.sprite = revealSprite;
            //如果该 Tile 四周没有地雷，则继续翻开它周围的所有 Tile
            if (MineCountNearby == 0)
            {
                foreach (Tile t in map.GetAdjacentTiles(this))
                {
                    t.Reveal();
                }//for
            }
            else
            {
                //如果周围有地雷，则显示周围地雷的数量
                mineCountText.gameObject.SetActive(true);
                mineCountText.text = MineCountNearby.ToString();
            }
        }//hasMine
        else
        {
            //游戏失败
            Debug.Log("Game Over!!!");
            //显示该 Tile 的地雷
            spriteRenderer.sprite = mineSprite;
        }
    }
}
```

④ 为 Tile 脚本新增的变量赋值。双击 TilePrefab 物体，为 Tile 脚本新增的变量赋值，如图 10-23 所示。

⑤ 单击"运行"按钮，用鼠标左键单击 Tile 可以翻开方格，如果周围无地雷，则会自动扩散，直到周围有地雷为止。翻开 Tile 的效果如图 10-24 所示。

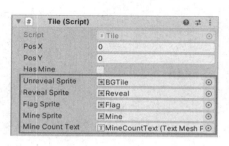

图 10-23　为 Tile 脚本新增的变量赋值

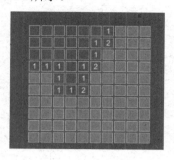

图 10-24　翻开 Tile 的效果

（2）右键标旗。

① 打开 Tile 脚本，修改 OnMouseOver()函数，当其检测到有右键单击 Tile 时，将当前 Tile 的状态更改为标旗/未翻开，并且当标识为旗子时，需要禁用左键。Tile 脚本中的代码如下：

```
using System.Collections;
using System.Collections.Generic;
using UnityEngine;

public class Tile : MonoBehaviour
{
    ……

    //鼠标指针在该Tile上
    void OnMouseOver()
    {
        //Debug.Log("Over " + posX + " " + posY);
        //左键按下后弹起
        if (Input.GetMouseButtonUp(0))
        {
            Reveal();        //翻开
        }
        //右键按下后弹起
        if (Input.GetMouseButtonUp(1))
        {
            //标旗或取消标旗
            MakeFlag();
        }
    }


    //翻开
    private void Reveal()
    {
        //如果被单击的Tile当前状态为翻开，则退出函数
        //如果被单击的Tile当前状态为标旗，则禁用左键
        if (state == State.Reveal || state == State.Flag) return;
```

```csharp
    ……
    }

    //标旗或取消标旗
    void MakeFlag()
    {
        //如果 Tile 为未翻开
        if (state == State.Unreveal)
        {
            state = State.Flag;
            spriteRenderer.sprite = flagSprite;
        }
        else if (state == State.Flag)
        {
            state = State.Unreveal;
            spriteRenderer.sprite = unrevealSprite;
        }
    }
}
```

② 返回 Unity 界面,单击"运行"按钮,此时用右键单击某个 Tile 便能为其标旗或取消标旗,并且当这个 Tile 为标旗状态时,翻开操作无效。标旗效果如图 10-25 所示。

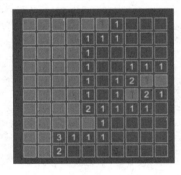

图 10-25 标旗效果

10.4.4 游戏结束判断

(1) 打开 Map 脚本,编写 WinCheck()函数、GameOver()函数等用于判断游戏是否结束。Map 脚本中的代码如下:

```csharp
using System.Collections;
using System.Collections.Generic;
using UnityEngine;

public class Map : MonoBehaviour
{
    ……

    //用于保存当前还有多少个 Tile 没有被翻开
    private int tilesLeftToReveal;
    //场景中已经标记的旗子数量
    private int flagCount;
```

......

```csharp
//初始化地雷
void GenerateMines()
{
    //设置需要标旗的数量
    flagCount = mineCount;
    //使用for循环逐个生成地雷
    for (int mineIndex = 0; mineIndex < mineCount; mineIndex++)
    {
        Tile tileWithMine;
        //随机遍历地图中的Tile，如果被遍历到的Tile已经有地雷，则重新遍历
        do
        {
            tileWithMine = map[Random.Range(0, rowNum),
                               Random.Range(0, colNum)];
        } while (tileWithMine.hasMine);
        //将遍历到的满足条件（没有地雷）的Tile设置为有地雷状态
        tileWithMine.AddMine();
        //把遍历到的有地雷的Tile放到地雷列表中
        tileWithMines.Add(tileWithMine);
        //需要被翻开的Tile的个数为总Tile数 - 地雷数
        tilesLeftToReveal = rowNum * colNum - mineCount;
    }
}
```

......

```csharp
//游戏胜利判断
public bool WinCheck()
{
    //如果未翻开的Tile的数量为0 并且旗子已经用完，则游戏胜利
    if (tilesLeftToReveal == 0 && flagCount == 0)
    {
        Debug.Log("You Win The Game!!!");
        return true;
    }
    return false;
}

//游戏失败操作
public bool GameOver(Sprite mineSprite)
{
    foreach (Tile t in tileWithMines)
    {
        t.GetComponent<SpriteRenderer>().sprite = mineSprite;
    }
    return true;
}

//每标记一面旗子，则旗子面数-1
public void DecreaseFlagNum()
```

```
        {
            flagCount--;
            Debug.Log("flagCount = " + flagCount);
        }

        //每取消标记一面旗子，则旗子面数+1
        public void IncreaseFlagNum()
        {
            flagCount++;
            Debug.Log("flagCount = " + flagCount);
        }

        //减少未翻开 Tile 的数量
        public void DescreaseTileToReveal()
        {
            tilesLeftToReveal--;
            Debug.Log("tilesLeftToReveal " + tilesLeftToReveal);
        }
}
```

（2）在 Tile 脚本中，需要在 MakeFlag()等函数中调用 WinCheck()函数。Tile 脚本中的代码如下：

```
using System.Collections;
using System.Collections.Generic;
using UnityEngine;

public class Tile : MonoBehaviour
{
    ……

    //翻开
    private void Reveal()
    {
        //如果被单击的 Tile 当前状态为翻开，则退出函数
        if (state == State.Reveal || state == State.Flag) return;
        //如果该 Tile 没有地雷，则可以执行以下操作
        if (hasMine == false)
        {
            state = State.Reveal;
            spriteRenderer.sprite = revealSprite;

            //如果翻开一个 Tile，则 Tile 未翻开的数量-1
            map.DescreaseTileToReveal();
            //判断游戏是否胜利
            if (map.WinCheck())
            {
                //出现游戏胜利的相关 UI
            }

            //如果该 Tile 四周没有地雷，则继续翻开它周围的所有 Tile
            if (MineCountNearby == 0)
            {
                foreach (Tile t in map.GetAdjacentTiles(this))
```

```csharp
                    {
                        t.Reveal();
                    }//for
                }
                else
                {
                    //如果周围有地雷，则显示周围地雷的数量
                    mineCountText.gameObject.SetActive(true);
                    mineCountText.text = MineCountNearby.ToString();
                }
            }//hasMine
            else
            {
                //游戏失败
                Debug.Log("Game Over!!!");
                map.GameOver(mineSprite);
                //显示该Tile的地雷
                spriteRenderer.sprite = mineSprite;
            }
        }


        //标旗或取消标旗
        void MakeFlag()
        {
            //如果Tile为未翻开
            if (state == State.Unreveal)
            {
                state = State.Flag;
                spriteRenderer.sprite = flagSprite;

                //剩余旗子面数-1
                map.DescreaseFlagNum();
                //判断游戏是否胜利
                if (map.WinCheck())
                {
                    //出现游戏胜利的相关UI
                }
            }
            else if (state == State.Flag)
            {
                state = State.Unreveal;
                spriteRenderer.sprite = unrevealSprite;

                //剩余旗子面数+1
                map.IncreaseFlagNum();
            }
        }
    }
```

（3）返回Unity界面，单击"运行"按钮，现在每标识一面旗子，Console窗口中便会实时地打印出flagCount的信息；当玩家踩到地雷时，Console窗口中会打印出玩家失败的信息；当玩家游戏胜利时，Console窗口中会打印出玩家胜利的信息，如图10-26所示。

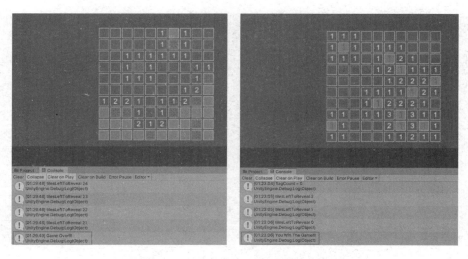

图 10-26　玩家失败/玩家胜利的信息

10.4.5　UI 控制

1．UI 制作

（1）将 Game 窗口中的画面比例改为 4∶3，这是为了使 UI 可以自适应，如图 10-27 所示。

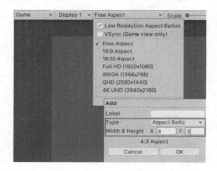

图 10-27　设置 Game 窗口中的画面比例

（2）调整 Panel 的大小。返回 Unity 界面，在主菜单栏中选择 GameObject→UI→Panel 命令，创建一个 UI 容器。把 4 个小三角形对准 Panel 的 4 个顶点（这 4 个小三角形的作用是控制 UI 随着屏幕大小进行缩放），如图 10-28 所示。

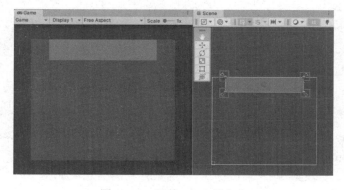

图 10-28　调整 Panel 的大小

（3）设置九宫格。把 Panel 的 Image 组件的 Source Image 设置为 top-bar，此时可能会发现 Panel 出现了边缘拉伸失真的问题，可以通过将 top-bar 设置为九宫格来解决这个问题。选中 top-bar，打开 Sprite Editor，可以看到图片的边缘有一圈绿色的线，如图 10-29 所示，将它们稍微拉往图片中心，单击 Apply 按钮保存。解决的原理简单来说就是这种九宫格的设置限制了图片的拉伸范围，从而解决了边缘拉伸失真的问题。采用同样的方法设置 Mines_left、time_left、face_holder 和 BGOhh。

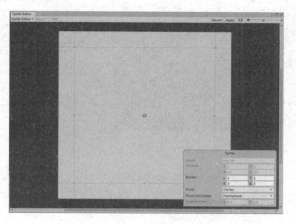

彩色图

图 10-29　设置九宫格

（4）修改 Image 组件的属性。将 Panel 的 Image 组件的 Image Type 设置为 Sliced，如图 10-30 所示。可以看到，图片边缘拉伸失真的问题已经解决。

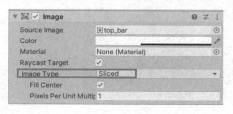

图 10-30　修改 Image 组件的属性

（5）设置 MineImage-UI。在 Panel 下新建 Image 并将其命名为 MineImage，将 Source Image 设置为 mines_left 图片，将 Image Type 设置为 Sliced，同时将 MineImage 图片摆放到合适的位置，如图 10-31 所示。

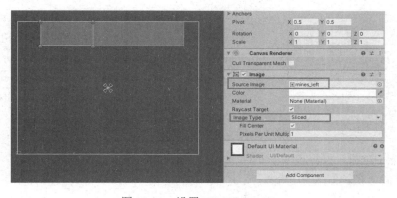

图 10-31　设置 MineImage-UI

（6）设置 FaceHolder-UI。采用同样的步骤将 face_holder 图片也摆放到合适的位置，并将其命名为 FaceHolder，如图 10-32 所示。

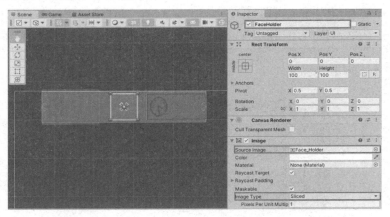

图 10-32　设置 FaceHolder-UI

（7）设置 TimerImage-UI。采用同样的步骤将 time_left 图片也摆放到合适的位置，并将其命名为 TimerImage，如图 10-33 所示。

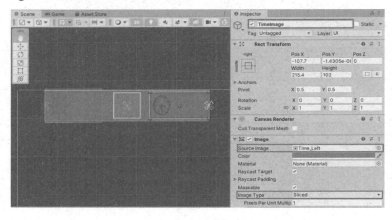

图 10-33　设置 TimerImage-UI

（8）设置 Canvas 属性。选中 Canvas，将其 Canvas Scaler 组件的 UI Scale Mode 设置为 Scale With Screen Size，如图 10-34 所示。

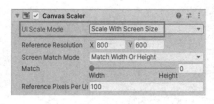

图 10-34　设置 Canvas 属性

（9）设置图片的中心点。选中 MineImage，按住 Alt 键，并单击其 Rect Transform 组件左上角的图标，将它的中心点改为左边中间，如图 10-35 所示。同样，将 FaceHolder 的中心移到正中间，将 TimerImage 的中心移到右边中间。在设置完成后，可以尝试缩放游戏画面，此时 UI 便会保持原来的比例和位置进行缩放。

（10）为图片添加文字信息子物体。在 MineImage 下新建一个 Text 对象，先将该对象的字号设置为 60，对齐方式设置为中间对齐，文字内容默认为空，再调整它在场景中的位置。按住 Alt 键单击其 Rect Transform 组件左上角的图标，将它的中心点改为右边中间。同样，对 TimerImage 也进行上述操作，如图 10-36 所示。

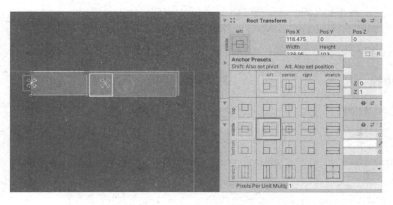

图 10-35　设置图片的中心点

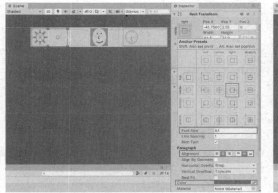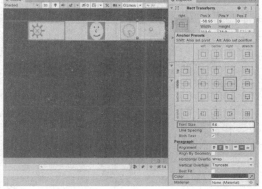

图 10-36　为图片添加文字信息子物体

（11）在 FaceHolder 下新建一个 Image 对象并将其命名为 Face，调整该对象的大小和位置。先为 Face 对象预设一个表情，如图 10-37 所示。

图 10-37　新建 Image 对象

（12）在制作完成后运行游戏，可以看到如图 10-38 所示的画面效果。

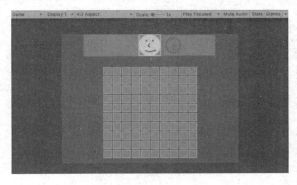

图 10-38　画面效果

2. 剩余地雷数

（1）在 Scripts 文件夹下新建一个文件夹并将其命名为 UIController，该文件夹用于保存控制 UI 的所有脚本。在 UIController 文件夹下新建一个脚本并将其命名为 MineLeftOverUI，该脚本专门用来控制剩余地雷数的显示。MineLeftOverUI 脚本中的代码如下：

```
using System.Collections;
using System.Collections.Generic;
using UnityEngine;
using TMPro;

//保证添加了该组件的对象已经有 TextMeshPro 组件
[RequireComponent(typeof(TextMeshProUGUI))]
public class MineLeftOverUI : MonoBehaviour
{
    public void SetMineLeftOver(int mineCount)
    {
        GetComponent<TextMeshProUGUI>().text = mineCount.ToString();
    }
}
```

（2）在 Map 脚本中引入 MineLeftOverUI 脚本，并调用 SetMineLeftOver() 函数。Map 脚本中的代码如下：

```
using System.Collections;
using System.Collections.Generic;
using UnityEngine;

public class Map : MonoBehaviour
{
    ……
    private MineLeftOverUI mineLeftOverText;

    void Start()
    {
        mineLeftOverText = FindObjectOfType<MineLeftOverUI>();
        CreateMap();         //生成地图
    }

    //初始化地雷
    void GenerateMines()
    {
        //设置需要标旗的数量
        flagCount = mineCount;
        //使用 for 循环逐个生成地雷
        for (int mineIndex = 0; mineIndex < mineCount; mineIndex++)
        {
            Tile tileWithMine;
            //随机遍历地图中的 Tile，
            //如果被遍历到的 Tile 已经有地雷，则重新遍历
            do
            {
                tileWithMine = map[Random.Range(0, rowNum),
                                    Random.Range(0, colNum)];
```

```
        } while (tileWithMine.hasMine);
        //将遍历到的满足条件(没有地雷)的Tile设置为有地雷状态
        tileWithMine.AddMine();
        //把遍历到的有地雷的Tile放入地雷列表中
        tileWithMines.Add(tileWithMine);

        //需要被翻开的Tile个数为总Tile数 - 地雷数
        tilesLeftToReveal = rowNum * colNum - mineCount;

        //显示地雷数
        mineLeftOverText.SetMineLeftOver(mineCount);
    }
}

......

//每标记一面旗子,则旗子面数-1
public void DecreaseFlagNum()
{
    flagCount--;
    Debug.Log("flagCount = " + flagCount);
    //显示剩余地雷数
    mineLeftOverText.SetMineLeftOver(flagCount);
}

//每取消标记一面旗子,则旗子面数+1
public void IncreaseFlagNum()
{
    flagCount++;
    Debug.Log("flagCount = " + flagCount);
    //显示剩余地雷数
    mineLeftOverText.SetMineLeftOver(flagCount);
}
}
```

（3）剩余地雷数的 UI 效果。返回 Unity 界面，将 MineLeftOverUI 脚本赋给 MineImage 下的 Text 子物体，单击"运行"按钮。每标记一面旗子，左上角剩余地雷的数字会相应地减 1；取消标记一面旗子，左上角剩余地雷的数字会相应地加 1，如图 10-39 所示。

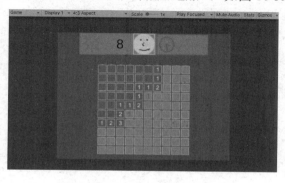

图 10-39　剩余地雷数的 UI 效果

3. 计时

（1）在 Scripts/UIController 文件夹下，新建一个脚本并将其命名为 TimeCountDownUI，

用来实现计时功能。TimeCountDownUI 脚本中的代码如下：

```csharp
using System.Collections;
using System.Collections.Generic;
using UnityEngine;
using TMPro;

[RequireComponent(typeof(TextMeshProUGUI))]
public class TimeCountDownUI : MonoBehaviour
{
    //暂停是否及时
    bool pause = true;
    //总时间
    private float timeCount;
    private TextMeshProUGUI text;
    //Start is called before the first frame update
    void Start()
    {
        text = GetComponent< TextMeshProUGUI>();
    }

    //Update is called once per frame
    void Update()
    {
        if (pause == true)
        {
            return;
        }
        else
        {
            //Time.deltaTime 为每两帧的时间间隔
            //因为 Update()函数是一帧调用一次，所以可以用 deltaTime 来累加时间
            timeCount += Time.deltaTime;
            //更新 UI
            UpdateText();
        }
    }

    void UpdateText()
    {
        //把时间赋值给 Text UI
        text.text = ((int)timeCount).ToString();
        //设置 text 的格式，让它显示成如 002、012、152 的格式。
        while (text.text.Length < 3)
        {
            text.text = "0" + text.text;
        }
    }

    //设置停止计时
    public void PauseTimer()
    {
        pause = true;
    }
```

```
    //设置开始计时
    public void StartTimer()
    {
        pause = false;
    }
    //获得计时器状态
    public bool IsPause()
    {
        return pause;
    }
}
```

（2）在 Map 脚本中新增 isGameStart 参数，用来表示游戏是否停止。Map 脚本中的代码如下：

```
using System.Collections;
using System.Collections.Generic;
using UnityEngine;

public class Map : MonoBehaviour
{
    ……

    //游戏停止
    private bool isGameStart = true;

    ……

    //游戏胜利的判断
    public bool WinCheck()
    {
        //如果未翻开 Tile 的数量为 0 并且旗子已经用完，则游戏胜利
        if (tilesLeftToReveal == 0 && flagCount == 0)
        {
            Debug.Log("You Win The Game!!!");
            isGameStart = false;
            return true;
        }
        return false;
    }

    //游戏失败操作
    public bool GameOver(Sprite mineSprite)
    {
        foreach (Tile t in tileWithMines)
        {
            t.GetComponent<SpriteRenderer>().sprite = mineSprite;
        }
        isGameStart = false;
        return true;
    }

    ……

    //获得游戏是否停止状态
```

```
    public bool IsGameStart()
    {
        return isGameStart;
    }
}
```

（3）在 Tile 脚本中引入 TimeCountDownUI 脚本，用来控制计时器是否停止计时。Tile 脚本中的代码如下：

```
using System.Collections;
using System.Collections.Generic;
using UnityEngine;

public class Tile : MonoBehaviour
{
    ……

    private TimeCountDownUI timeCountDownUI;

    private void Awake()
    {
        if (map == null)
        {
            map = GameObject.Find("Map").GetComponent<Map>();
        }
        spriteRenderer = GetComponent<SpriteRenderer>();
        timeCountDownUI = FindObjectOfType<TimeCountDownUI>();
    }

    void Update()
    {
        //如果游戏结束，则停止计时
        if (map.IsGameStart() == false)
        {
            timeCountDownUI.PauseTimer();
        }
    }

    //鼠标在该 Tile 上
    void OnMouseOver()
    {
        //Debug.Log("Over " + posX + " " + posY);

        if (map.IsGameStart())
        {
            //左键按下后弹起
            if (Input.GetMouseButtonUp(0))
            {
                if (timeCountDownUI.IsPause() == true)
                {
                    timeCountDownUI.StartTimer();
                }
                Reveal();        //翻开
            }
            //右键按下后弹起
            if (Input.GetMouseButtonUp(1))
```

```
        {
            if (timeCountDownUI.IsPause() == true)
            {
                timeCountDownUI.StartTimer();
            }
            //标旗
            MakeFlag();
        }
    }
    ……
}
```

（4）测试计时 UI。返回 Unity 界面，将 TimeCountDownUI 脚本附加给 TimerImage 的子物体 Text，单击"运行"按钮。在第一次单击任意一个方格或标记旗子时，计时器就开始计时，当游戏失败或胜利时，计时结束，如图 10-40 所示。

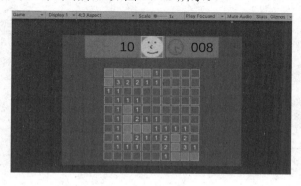

图 10-40　测试计时 UI

4．表情控制

（1）在 Scripts/UIController 文件夹下，新建一个脚本并将其命名为 FaceController，用于控制表情 UI。FaceController 脚本中的代码如下：

```
using System.Collections;
using System.Collections.Generic;
using UnityEngine;
using UnityEngine.UI;

public enum Face_State { Ok, Ohh, Dead, Wow }
[RequireComponent(typeof(Image))]
public class FaceController : MonoBehaviour
{

    Face_State state;
    public Sprite OkSprite;
    public Sprite OhhSprite;
    public Sprite DeadSprite;
    public Sprite WowSprite;
    private Image faceImage;
    //Start is called before the first frame update
    void Start()
    {
```

```
        faceImage = GetComponent<Image>();
    }

    //Update is called once per frame
    void Update()
    {

    }

    public void SetState(Face_State state)
    {
        this.state = state;
        switch (state)
        {
            case Face_State.Ok:
                faceImage.sprite = OkSprite;
                break;
            case Face_State.Ohh:
                faceImage.sprite = OhhSprite;
                StartCoroutine(ReturnToOkState());
                break;
            case Face_State.Dead:
                faceImage.sprite = DeadSprite;
                break;
            case Face_State.Wow:
                faceImage.sprite = WowSprite;
                break;
        }
    }

    //一段时间后回到OK图片
    IEnumerator ReturnToOkState()
    {
        yield return new WaitForSeconds(0.1f);
        if (faceImage.sprite == OhhSprite)
        {
            faceImage.sprite = OkSprite;
            SetState(Face_State.Ok);
        }
    }
}
```

（2）在 Tile 脚本中引入 FaceController 脚本，用于实现对表情的控制。Tile 脚本中的代码如下：

```
using System.Collections;
using System.Collections.Generic;
using UnityEngine;

public class Tile : MonoBehaviour
{
    ……

    public FaceController faceController;
```

```csharp
private void Awake()
{
    if (map == null)
    {
        map = GameObject.Find("Map").GetComponent<Map>();
    }
    spriteRenderer = GetComponent<SpriteRenderer>();
    timeCountDownUI = FindObjectOfType<TimeCountDownUI>();
    faceController = FindObjectOfType<FaceController>();
}

//鼠标指针在该Tile上
void OnMouseOver()
{
    //Debug.Log("Over " + posX + " " + posY);

    if (map.IsGameStart())
    {
        //左键按下后弹起
        if (Input.GetMouseButtonUp(0))
        {
            faceController.SetState(Face_State.Ohh);

            if (timeCountDownUI.IsPause() == true)
            {
                timeCountDownUI.StartTimer();
            }

            Reveal();         //翻开Tile
        }
        //右键按下后弹起
        if (Input.GetMouseButtonUp(1))
        {
            if (timeCountDownUI.IsPause() == true)
            {
                timeCountDownUI.StartTimer();
            }
            faceController.SetState(Face_State.Ok);
            //标旗
            MakeFlag();
        }
    }
}

//翻开Tile
private void Reveal()
{
    //如果被单击Tile的当前状态为翻开,则退出函数
    if (state == State.Reveal || state == State.Flag) return;
    //如果该Tile没有地雷,则可以执行以下操作
    if (hasMine == false)
    {
```

```csharp
            state = State.Reveal;
            spriteRenderer.sprite = revealSprite;

            //如果翻开一个Tile,则未翻开Tile的数量-1
            map.DescreaseTileToReveal();
            //判断游戏是否胜利
            if (map.WinCheck())
            {
                faceController.SetState(Face_State.Wow);
                timeCountDownUI.PauseTimer();
            }
            //如果该Tile四周没有地雷,则继续翻开它周围的所有Tile
            if (MineCountNearby == 0)
            {
                foreach (Tile t in map.GetAdjacentTiles(this))
                {
                    t.Reveal();
                }//for
            }
            else
            {
                //如果周围有地雷,则显示出周围地雷的数量
                mineCountText.gameObject.SetActive(true);
                mineCountText.text = MineCountNearby.ToString();
            }
        }//hasMine
        else
        {
            //游戏失败
            Debug.Log("Game Over!!!");
            //显示该Tile的地雷
            if (map.GameOver(mineSprite))
            {
                //设置OK图片
                faceController.SetState(Face_State.Dead);
                timeCountDownUI.PauseTimer();
            }
        }
    }

    //标旗或取消标旗
    void MakeFlag()
    {
        //如果Tile为未翻开
        if (state == State.Unreveal)
        {
            state = State.Flag;
            spriteRenderer.sprite = flagSprite;
            //剩余旗子面数-1
            map.DescreaseFlagNum();
            //判断游戏是否胜利
            if (map.WinCheck())
            {
```

```
            //出现游戏胜利的相关UI
            faceController.SetState(Face_State.Wow);
        }
    }
    else if (state == State.Flag)
    {
        state = State.Unreveal;
        spriteRenderer.sprite = unrevealSprite;
        //剩余旗子面数+1
        map.IncreaseFlagNum();
    }
}
```

（3）返回 Unity 界面，先将 FaceController 脚本赋值给 FaceHolder 的子物体 Face，再为 FaceController 脚本中的图片变量赋值，如图 10-41 所示。

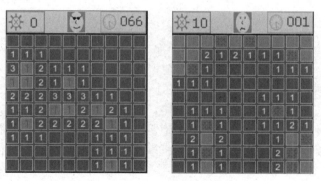

图 10-41　为 FaceController 脚本中的图片变量赋值

（4）测试游戏胜利/失败时的表情。单击"运行"按钮，测试表情控制是否正确，如图 10-42 所示。

图 10-42　测试游戏胜利/失败时的表情

10.4.6　重新开始一局游戏

还需要添加最后一项功能，当按下 Esc 键时，重新开始一局游戏。

（1）在 Map 脚本中导入 UnityEngine.SceneManagement 包，在 Update()函数中实现重新开始一局游戏的功能。Map 脚本中的代码如下：

```
using System.Collections;
using System.Collections.Generic;
using UnityEngine;
using UnityEngine.SceneManagement;


public class Map : MonoBehaviour
```

```
    {
        ……

        void Update()
        {
            //按下 Esc 键后重新开始一局游戏
            if (Input.GetKeyUp(KeyCode.Escape))
            {
                SceneManager.LoadScene(0);
            }
        }
        ……
    }
```

（2）返回 Unity 界面，单击"运行"按钮。测试按下 Esc 键是否能成功重新载入游戏场景，开始一局新游戏。

10.5　本章小结

本章主要介绍了《扫雷》游戏的设计与实现，其中重点在于方格的数据结构设计、地雷的随机分布，以及单击方格后的响应判定等。

10.6　练习题

1．在《扫雷》游戏中，假设在玩家第一次单击 $m×n$ 阵列中的某个方格之前就已经随机排列布好 k 个地雷，且在任何一个方格出现地雷的概率均相等，该概率为多少？

2．在本章介绍的《扫雷》游戏中，生成地雷后必须在非地雷方格上标记周围的地雷数。标记地雷采用的方法如下：遍历所有地雷方格，以每个地雷方格为中心，更新其周围方格的信息。现考虑另一种实现方法：遍历所有非地雷方格，以每个非地雷方格为中心，统计其周围的地雷数，并更新中心方格的信息。试分析这两种实现方法在效率上的区别，哪一种方法更好？

3．（选做）在本章介绍的《扫雷》游戏中，当玩家单击到空白方格（周围无地雷）时，会自动递归展开，直至遇到非空白方格（周围有地雷），从而形成一个以非空白方格为边界线的连通域。现考虑另一种循环实现方法（种子填充算法）：首先从当前空白方格分别向左、右两侧试探，直至找到非空白方格为止（类似于《连连看》游戏中的三连判断）；然后分别对已试探过的方格的上、下两个方格同样执行上述操作，直至上、下方向也遇到非空白方格为止（为了提升效率，应记录已试探过的方格，使其不再被试探）。试分析这两种实现方法在效率上的区别，哪一种方法更好？

第 11 章

贪吃蛇

11.1 游戏简介

《贪吃蛇》游戏最早的原型可以追溯到 1976 年。该游戏最初是在街机上运行的，原名为 *Blockade*。*Blockade* 是一款双人游戏，如图 11-1 所示，由 Gremlin 作为发行商发行。在 *Blockade* 游戏的街机版本中，蛇不会向前移动，只会尾巴不动、头部逐渐变长。实际上，它的设定并不是蛇，而是两个小人一边向前走一边在身后筑墙。*Blockade* 游戏的规则是，谁先撞到墙壁谁就输了。在个人计算机普及之前，已知的个人计算机版本是 TRS-80 型计算机上的 Worm 程序，该程序的作者是 Peter Trefonas。Worm 程序在当时的界面如图 11-2 所示。《贪吃蛇》游戏之所以被人们熟知，主要源于其是诺基亚手机内置的一款游戏，如图 11-3 所示。

 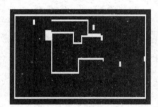 

图 11-1　*Blockade*　　图 11-2　Worm 程序在当时的界面　　图 11-3　诺基亚手机上的《贪吃蛇》游戏

11.2 游戏规则

游戏地图是一个长方形场地，可以视为一个由 30×20 个方格组成的地图，食物和组成蛇身的块均占一个小方格。玩家操控一条贪吃蛇在长方形场地中行走，贪吃蛇会按照玩家所按的方向行走或转弯。当贪吃蛇吃到食物后，蛇身会变长，并得分；当贪吃蛇撞到墙壁或自身时，游戏结束或损失一条生命。

11.3 游戏开发核心思路

11.3.1 地图的生成

观察《贪吃蛇》游戏的场景，可以自然地联想到使用二维数组表达游戏场景。食物和组成蛇的部分各占一个元素单位。

11.3.2 食物的出现

由于食物在场景中是随机生成的，因此可以采用随机数生成食物出现的坐标。此时需要注意生成的位置是否在蛇体内部，即判断食物出现的坐标是否在贪吃蛇体内，具体就是将食物的坐标与贪吃蛇体内每个元素的坐标进行比较。若二者重合，则重新生成随机数，直到食物不在贪吃蛇体内为止。

11.3.3 蛇的数据结构

蛇的数据结构可变长，所以可以考虑使用链表存储蛇身所有的方格。当贪吃蛇吃到食物后，便在链表中添加一个元素。

11.3.4 贪吃蛇的移动算法

通过用户输入可以获得贪吃蛇的移动方向。如果用坐标的自加或自减运算来做贪吃蛇的移动，代码计算量会很大。

可以用一种简单的方法实现贪吃蛇的移动。当贪吃蛇没有吃到食物时，用一个变量存储蛇头移动前的位置；在蛇头移动前的位置添加一个元素，并移除尾部最后一个元素，实现贪吃蛇的移动，如图11-4所示。

当贪吃蛇吃到食物时，用一个变量存储蛇头移动前的位置，在蛇头移动前的位置添加新的方格，这样便实现了贪吃蛇的变长，如图11-5所示。

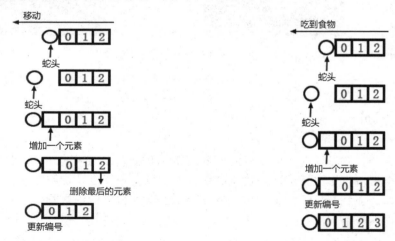

图 11-4　贪吃蛇的移动（没有吃到食物）　　图 11-5　贪吃蛇的移动（吃到食物）

11.3.5 判断蛇头是否撞到了自身

遍历组成贪吃蛇的所有的方格，判断非蛇头的位置是否与蛇头的位置相同。如果有相同的位置，则说明贪吃蛇撞到了自身；如果没有相同的位置，则说明贪吃蛇没有撞到自身。

11.3.6 边界的判断

将蛇头的坐标与边界的坐标进行比较。若蛇头的坐标与边界有重合的地方，则说明蛇头撞到了边界，游戏结束。

11.3.7 游戏流程图

游戏流程图如图 11-6 所示。

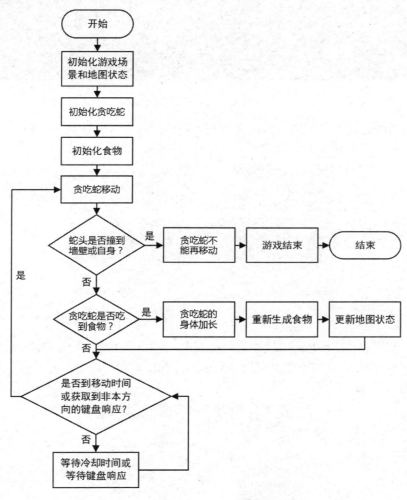

图 11-6 游戏流程图

11.4 游戏实现

11.4.1 资源导入和场景设置

（1）新建工程。如图 11-7 所示，选择 2D 选项，新建一个名为 Snake 的工程。

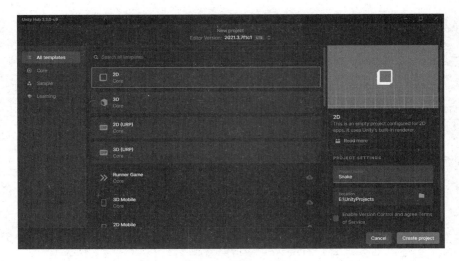

图 11-7　新建工程

（2）导入资源包。首先创建不同的文件夹，将不同的资源保存到不同的文件夹中；然后将下载的图片资源保存到 Sprites 文件夹中，如图 11-8 所示。

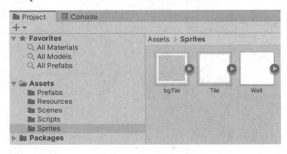

图 11-8　导入资源包

（3）修改图片的 Pixels Per Unit 参数。在 Sprites 文件夹中选择 bgTile 图片、Tile 图片和 Wall 图片，在 Inspector 窗口中调整它们 Pixels Per Unit 参数，将图片大小与 Unity 中的单位对应，如图 11-9 所示。

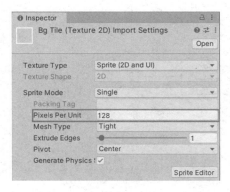

图 11-9　调整图片的 Pixels Per Unit 参数

（4）调整摄像机的位置和视野。在 Hierarchy 窗口中单击 Main Camera 物体后，在 Inspector 窗口中调整 Position 参数和 Size 参数，如图 11-10 所示。

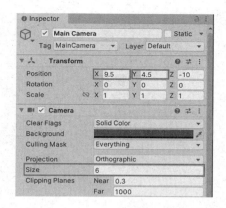

图 11-10　调整摄像机的位置和视野

11.4.2　编写 Node 类

（1）新建一个名为 Node 的脚本，该脚本用于设置节点，利用枚举类型来判断 Node 对象的状态，以及加载各种图片。Node 脚本中的代码如下：

```
using System.Collections;
using System.Collections.Generic;
using UnityEngine;

public enum NodeState{
    //Node 对象的状态类型
    Blank = 0, Wall = -1, Snake = 1,Food = 2
}
public class Node : MonoBehaviour
{
    public int x{private set;get;}              //该 Node 对象的整型 x 坐标，
                                                //其中的花括号标记该变量只能在该类中设置，
                                                //不能被其他类设置，但可以被其他类获得数值
    public int y{private set;get;}              //该 Node 对象的整型 y 坐标
    NodeState nodeState;                        //该 Node 对象的状态
    private SpriteRenderer mapTileSR;
    public Sprite blankSprite;                  //空的图片
    public Sprite wallSprite;                   //墙壁的图片
    public Sprite snakeSprite;                  //贪吃蛇的图片
    public Sprite foodSprite;                   //食物的图片

    void Awake(){
        mapTileSR = GetComponent<SpriteRenderer>(); //获得 Sprite Renderer 组件
    }

    //设置该 Node 对象的整型位置，以及在场景中的位置
    public void SetNodePos(int x, int y){
        this.x = x;
        this.y = y;
        transform.position = new Vector3(x,y,0);
    }
    //设置该 Node 对象的状态类型
    public void SetNodeState(NodeState nodeState){
```

```
    //设置该 Node 对象的状态
    this.nodeState = nodeState;
    //根据该 Node 对象的状态为该 Node 对象设置图片及颜色
    switch(nodeState){
        case NodeState.Blank:
            mapTileSR.sprite = blankSprite;
            mapTileSR.color = Color.black;
        break;
        case NodeState.Wall:
            mapTileSR.sprite = wallSprite;
            mapTileSR.color = Color.gray;
        break;
        case NodeState.Snake:
            mapTileSR.sprite = snakeSprite;
            mapTileSR.color = Color.white;
        break;
        case NodeState.Food:
            mapTileSR.sprite = foodSprite;
            mapTileSR.color = Color.blue;
        break;
    }
}
//获得该 Node 对象的状态
public NodeState GetNodeState(){
    return nodeState;
}
}
```

（2）在场景中创建一个空对象并将其命名为 Node。先添加 Sprite Renderer 组件和 Node 脚本，再为 Node 脚本中的变量赋值，并且把 Node 对象拖入 Prefabs 文件夹中保存为预制体，如图 11-11 所示。

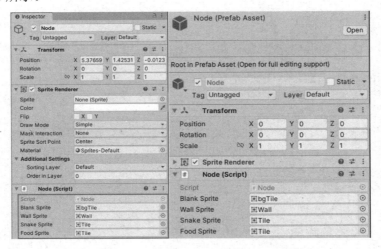

图 11-11　为 Node 对象添加脚本、赋值，并保存为预制体

11.4.3　初始化场景

（1）通过二维数组来表达场景。在 Scripts 文件夹下新建一个脚本并将其命名为

GameController。GameController 脚本中的代码如下：

```csharp
using System.Collections;
using System.Collections.Generic;
using UnityEngine;

public class GameController : MonoBehaviour
{
    Node[,] snapshotMap;                                //地图快照
    public int rowNum = 10;                             //地图总行数
    public int colNum = 20;                             //地图总列数
    public Node nodePrefab;                             //Node 预制体

    //Start is called before the first frame update
    void Start()
    {
        InitMap();
    }

    void InitMap(){
        //1.创建场景
        CreateBackground();
    }

    void CreateBackground(){
        snapshotMap = new Node[rowNum,colNum];          //为快照分配内存空间
        for(int row = 0; row < rowNum; row++){
            for(int col = 0; col < colNum; col++){
                Node tempNode = Instantiate(nodePrefab);    //生成一个 Node 对象
                tempNode.SetNodePos(col,row);               //设置该 Node 对象的位置
                tempNode.transform.SetParent(transform);    //整理 Hierarchy 窗口中的对象
                snapshotMap[row,col] = tempNode;            //把该 Node 对象放入快照中
                if(row == 0 || row == rowNum - 1 || col == 0 || col == colNum - 1){
                    snapshotMap[row,col].SetNodeState(NodeState.Wall); //设置墙壁
                }else{
                    //设置为空的 Node 对象
                    snapshotMap[row,col].SetNodeState(NodeState.Blank);
                }
            }//end col
        }//end row
    }
}
```

（2）在场景中创建一个空物体并将其命名为 GameController，将该物体的 Position 参数的值归零，如图 11-12 所示；挂载 GameController 脚本并赋值，如图 11-13 所示。

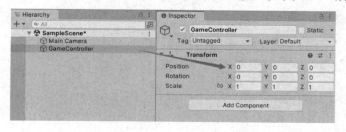

图 11-12　将 GameController 物体的 Position 参数的值归零

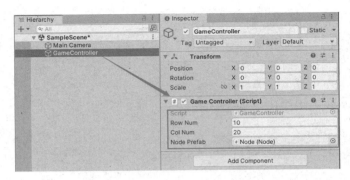

图 11-13　挂载 GameController 脚本并赋值

（3）运行效果如图 11-14 所示。

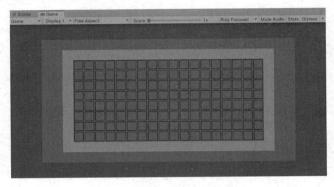

图 11-14　运行效果

（4）创建贪吃蛇和食物。在 Unity 中创建一个 Snake 类的脚本，用于控制贪吃蛇的移动；创建一个名为 Snake 的空物体并挂载 Snake 脚本，关闭 Snake 物体的 enable 属性。图 11-15 所示为为 Snake 物体挂载脚本和关闭 enable 属性。

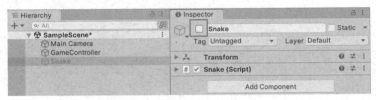

图 11-15　为 Snake 物体挂载脚本和关闭 enable 属性

（5）设置贪吃蛇和食物的生成位置，并且刷新快照。在 GameController 脚本中添加新代码（加粗字体为在原有代码的基础上新增的内容）。GameController 脚本中的代码如下：

```
public class GameController : MonoBehaviour
{
    Node[,] snapshotMap;                    //地图快照
    public int rowNum = 10;                 //地图总行数
    public int colNum = 20;                 //地图总列数
    public Node nodePrefab;                 //Node 预制体

    public Snake snake;                     //贪吃蛇对象

    int foodRow,foodCol;                    //食物的位置
```

```csharp
//Start is called before the first frame update
void Start()
{
    InitMap();
}
void InitMap(){
    //1.创建场景
    CreateBackground();
    //2.创建贪吃蛇
    CreateSnake();
    //3.创建食物
    CreateFood();
}

void CreateBackground(){
    snapshotMap = new Node[rowNum,colNum];                  //为快照分配内存空间

    for(int row = 0; row < rowNum; row++){
        for(int col = 0; col < colNum; col++){
            Node tempNode = Instantiate(nodePrefab);        //生成一个 Node 对象
            tempNode.SetNodePos(col,row);                   //设置该 Node 对象的位置
            tempNode.transform.SetParent(transform);        //整理 Hierarchy 窗口中的对象
            snapshotMap[row,col] = tempNode;                //把该 Node 对象放入快照中
            if(row == 0 || row == rowNum - 1 || col == 0 || col == colNum - 1){
                snapshotMap[row,col].SetNodeState(NodeState.Wall);   //设置墙壁
            }else{
                //设置为空的 Node 对象
                snapshotMap[row,col].SetNodeState(NodeState.Blank);
            }
        }//end col
    }//end row
}

//创建贪吃蛇
void CreateSnake(){
    snake.gameObject.SetActive(true);
    snake.CreateSnake((int)(colNum*0.5),(int)(rowNum*0.5));
}
//创建食物
public void CreateFood(){
    do{
        foodRow = Random.Range(1,rowNum-1);
        foodCol = Random.Range(1,colNum-1);
        Debug.Log("foodRow = " + foodRow + " foodCol = " + foodCol);
    }while(snapshotMap[foodRow,foodCol].GetNodeState() != NodeState.Blank);
    //刷新快照
    UpdateMap();
}

//刷新快照
public void UpdateMap(){
    for(int row = 1; row < rowNum-1;row++){
        for(int col = 1; col < colNum-1;col++){
```

```
            //1.清空快照
            snapshotMap[row,col].SetNodeState(NodeState.Blank);
            //2.设置食物在快照中的位置
            if(row == foodRow && col == foodCol){
                snapshotMap[row,col].SetNodeState(NodeState.Food);
            }
        }//col
    }//row
    //3.设置贪吃蛇在快照中的位置
}
```

(6) 为 GameController 脚本中的 Snake 物体赋值, 如图 11-16 所示。

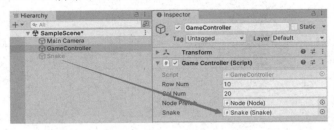

图 11-16 为 GameController 脚本中的 Snake 物体赋值

(7) 运行程序。可以看到地图上随机生成了食物, 如图 11-17 所示。

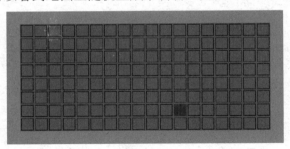

图 11-17 生成食物

11.4.4 贪吃蛇的移动

(1) 在 Snake 脚本中用 List 来表示贪吃蛇的身体。在 Snake 脚本中添加 GetSnakeLength() 函数和 GetSnakeTilePos()函数, 用于获取贪吃蛇的长度和贪吃蛇的每个小节在快照中的位置。其中, Add()函数的作用是向 List 中添加一个数。Snake 脚本中的代码如下:

```
public enum DirState{Left,Right,Up,Down};          //贪吃蛇的移动方向
public class Snake : MonoBehaviour
{
    private List<Vector2> snake;                    //贪吃蛇的数据结构
    //创建贪吃蛇
    public void CreateSnake(int x, int y){
        snake = new List<Vector2>();                //创建贪吃蛇链表
        Vector2 head = new Vector2(x,y);            //创建贪吃蛇的头
        snake.Add(head);                            //把蛇头加入链表中
    }
    //获得贪吃蛇的长度
```

```
    public int GetSnakeLength(){
        return snake.Count;
    }
    //获得贪吃蛇的每个小节的位置
    public Vector2 GetSnakeTilePos(int index){
        return snake[index];
    }
}
```

（2）返回 GameController 脚本，调用 Snake 脚本中的函数用于创造贪吃蛇并在快照中更新。GameController 脚本中的代码如下（省略号表示之前未修改的函数，此处不再赘述）：

```
public class GameController : MonoBehaviour
{
    ……
    //创建贪吃蛇
    void CreateSnake(){
        snake.gameObject.SetActive(true);
        snake.CreateSnake((int)(colNum*0.5),(int)(rowNum*0.5));
    }
    ……
    //刷新快照
    public void UpdateMap(){
        for(int row = 1; row < rowNum-1;row++){
            for(int col = 1; col < colNum-1;col++){
                //1.清空快照
                snapshotMap[row,col].SetNodeState(NodeState.Blank);
                //2.设置食物在快照中的位置
                if(row == foodRow && col == foodCol){
                    snapshotMap[row,col].SetNodeState(NodeState.Food);
                }
            }//col
        }//row
        //3.设置贪吃蛇在快照中的位置
        for(int index = 0; index < snake.GetSnakeLength();index++){
            Vector2 snakeTilePos = snake.GetSnakeTilePos(index);
            snapshotMap[(int)snakeTilePos.y,(int)snakeTilePos.x].SetNodeState
                                                    (NodeState.Snake);
        }
    }
}
```

（3）测试贪吃蛇的移动效果。运行游戏，可以看到蛇头在地图上显现，如图 11-18 所示。

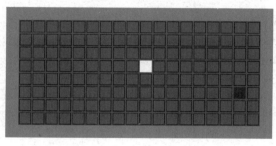

图 11-18　测试贪吃蛇的移动效果

(4) 在 Snake 脚本中使用枚举类型来表达贪吃蛇的移动方向。设置贪吃蛇的移动速度,并添加 DirDetection()函数用于监听键盘操作。Snake 脚本中的代码如下:

```csharp
public enum DirState{Left,Right,Up,Down};            //贪吃蛇的移动方向
public class Snake : MonoBehaviour
{
    private List<Vector2> snake;                     //贪吃蛇的数据结构
    DirState dirState = DirState.Left;               //贪吃蛇移动方向的状态
    private Vector2 moveDir = Vector2.left;          //贪吃蛇移动方向的向量数值
    public float moveSpeed = 0.3f;
    public GameController gameController;

    //Start is called before the first frame update
    void Start()
    {
        gameController = GameObject.FindObjectOfType<GameController>();
    }
    //Update is called once per frame
    void Update()
    {
        DirDetection();
    }
    ......
    //监听键盘操作,判断贪吃蛇的移动方向
    void DirDetection(){
        //向下
        if(Input.GetKeyDown(KeyCode.DownArrow) && dirState != DirState.Up){
            dirState = DirState.Down;
            moveDir = Vector2.down;
        }
        //向上
        if(Input.GetKeyDown(KeyCode.UpArrow) && dirState != DirState.Down){
            dirState = DirState.Up;
            moveDir = Vector2.up;
        }
        //向左
        if(Input.GetKeyDown(KeyCode.LeftArrow) && dirState != DirState.Right){
            dirState = DirState.Left;
            moveDir = Vector2.left;
        }
        //向右
        if(Input.GetKeyDown(KeyCode.RightArrow) && dirState != DirState.Left){
            dirState = DirState.Right;
            moveDir = Vector2.right;
        }
    }
    ......
}
```

(5) 为了使贪吃蛇移动,需要在 Snake 脚本中添加 Move()函数。其中,Insert 方法用于对 List 插入一个值,第一个参数为在 List 中的位置,第二个参数为插入的值。RemoveAt 方法用于移除 List 中的一个值,其中的参数为将被移除数的位置。在 InvokeRepeating 方法中,第一

个参数为方法名,第二个参数为多少秒后执行,第三个参数为重复执行的时间间隔。

Snake 脚本中的代码如下:

```csharp
public enum DirState{Left,Right,Up,Down};              //贪吃蛇的移动方向
public class Snake : MonoBehaviour
{
    private List<Vector2> snake;                        //贪吃蛇的数据结构
    DirState dirState = DirState.Left;                  //贪吃蛇移动方向的状态
    private Vector2 moveDir = Vector2.left;             //贪吃蛇移动方向的向量数值
    public float moveSpeed = 0.3f;
    public GameController gameController;

    //Start is called before the first frame update
    void Start()
    {
        gameController = GameObject.FindObjectOfType<GameController>();
        InvokeRepeating("Move",0.3f,moveSpeed);
    }
    ……
    void Move(){
        Vector2 tempHeadPos = snake[0];                 //获得蛇头的位置
        tempHeadPos += moveDir;                         //把蛇头向移动方向移动一格
        //移动整条贪吃蛇
        snake[snake.Count-1] = tempHeadPos;
        //将贪吃蛇的最后一个元素作为中间变量,保存移动后的蛇头,相当于最后一个元素数据被清除
        snake.Insert(0,snake[snake.Count-1]);           //把蛇头放入第零个元素的位置
        snake.RemoveAt(snake.Count-1);                  //删除最后一个元素
        gameController.UpdateMap();                     //更新快照
    }
    ……
```

(6)通过键盘控制贪吃蛇的移动方向。运行游戏,按键盘上的方向键,测试控制贪吃蛇的移动方向的功能,如图 11-19 所示。

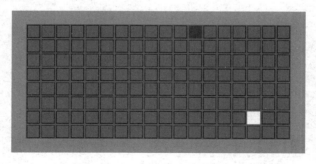

图 11-19　通过键盘控制贪吃蛇的移动方向

(7)贪吃蛇吃到食物。在 GameController 类中编写 3 个函数,分别为 SetMapItemState()、GetMapItemState()和 GetFoodPos(),用于获得快照上的信息。GameController 脚本中的代码如下:

```csharp
public class GameController : MonoBehaviour
{
    ……
```

```csharp
//设置当前位置的状态
public void SetMapItemState(int row,int col,NodeState nodeState){
    snapshotMap[row,col].SetNodeState(nodeState);
}
//获得当前位置的状态
public NodeState GetMapItemState(int row,int col){
    return snapshotMap[row,col].GetNodeState();
}

//获得食物的位置
public Vector2 GetFoodPos(){
    return new Vector2(foodCol,foodRow);
}
```

（8）贪吃蛇吃到食物。在 Snake 脚本中增加 EatFood 脚本，用于判断贪吃蛇是否吃到食物。如果吃到食物，则增加贪吃蛇的长度。Snake 脚本中的代码如下：

```csharp
public enum DirState{Left,Right,Up,Down};          //贪吃蛇的移动方向
public class Snake : MonoBehaviour
{
    ……
    void Move(){
        Vector2 tempHeadPos = snake[0];             //获得蛇头的位置
        tempHeadPos += moveDir;                     //把蛇头向移动方向移动一格
        //判断是否吃到食物
        if(EatFood((int)tempHeadPos.x,(int)tempHeadPos.y)){
            return;
        }else{
            //移动整条贪吃蛇
            snake[snake.Count-1] = tempHeadPos;
            //将贪吃蛇的最后一个元素作为中间变量，保存移动后的蛇头，相当于最后一个元素数据被清除
            snake.Insert(0,snake[snake.Count-1]);   //把蛇头放入第零个元素的位置
            snake.RemoveAt(snake.Count-1);          //删除最后一个元素
            gameController.UpdateMap();             //更新快照
        }
    }
    public bool EatFood(int x, int y){
        //判断贪吃蛇是否吃到食物
        if(gameController.GetMapItemState((int)y,(int)x) == NodeState.Food){
            //把该位置设置为空地
            gameController.SetMapItemState((int)y,(int)x,NodeState.Blank);
            //在食物的位置添加到蛇头，相当于增加了贪吃蛇的长度
            snake.Insert(0,gameController.GetFoodPos());
            //重新生成食物
            gameController.CreateFood();
            //更新地图
            gameController.UpdateMap();
            Debug.Log("Ate");
            return true;
        }
        return false;
    }
    ……
}
```

（9）测试贪吃蛇吃到食物的功能。运行游戏，当蛇头撞到食物时，食物会消失，并生成下一个食物，如图 11-20 所示。

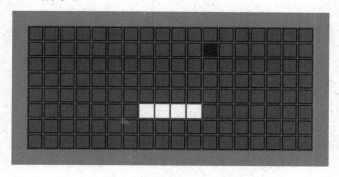

图 11-20　贪吃蛇吃到食物的效果

11.4.5　结束判定

（1）检测蛇头与墙壁或自身是否发生碰撞。若发生碰撞，则游戏结束。在 Snake 脚本中增加判断条件。CancelInvoke()函数用来终止 InvokeRepeating()函数的运行。Snake 脚本中的代码如下：

```csharp
public enum DirState{Left,Right,Up,Down};      //贪吃蛇的移动方向
public class Snake : MonoBehaviour
{
    void Move(){
        Vector2 tempHeadPos = snake[0];         //获得蛇头的位置
        tempHeadPos += moveDir;                 //把蛇头向移动方向移动一格
        //判断蛇头是否撞到墙壁或自身
        NodeState nodeState = gameController.GetMapItemState((int)tempHeadPos.y,
                                                             (int)tempHeadPos.x);
        if(nodeState == NodeState.Wall ||nodeState == NodeState.Snake){
            //终止贪吃蛇的移动
            CancelInvoke();
            Debug.Log("GameOver");
            return;
        }
        //判断贪吃蛇是否吃到食物
        if(EatFood((int)tempHeadPos.x,(int)tempHeadPos.y)){
            return;
        }else{
        //移动整条贪吃蛇
        snake[snake.Count-1] = tempHeadPos;
        //将贪吃蛇的最后一个元素作为中间变量，保存移动后的蛇头，相当于最后一个元素数据被清除
        snake.Insert(0,snake[snake.Count-1]);   //把蛇头放入第零个元素的位置
        snake.RemoveAt(snake.Count-1);          //删除最后一个元素
        gameController.UpdateMap();             //更新快照
    }
}
```

（2）测试游戏结束功能。运行游戏，蛇头撞到墙壁，游戏结束，如图 11-21 所示。

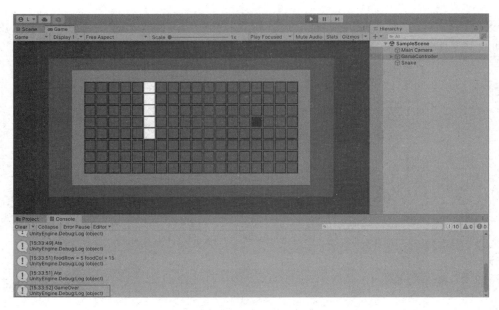

图 11-21　游戏结束

11.5　本章小结

本章主要介绍了《贪吃蛇》游戏的设计与实现，其中重点在于贪吃蛇的数据结构设计、贪吃蛇的移动算法、贪吃蛇的增长、蛇头的碰撞检测等。

11.6　练习题

1．本章介绍的《贪吃蛇》游戏的蛇头在撞到墙壁时会被判定为失败，现考虑一种允许贪吃蛇穿墙的设定，即允许贪吃蛇在撞到某面墙时可"钻入"该面墙，并从对面墙中"钻出"，从而可以继续保持同方向移动。例如，蛇头撞到左侧墙壁后从右侧墙壁同一行的位置出来，并继续向左前进。试实现该扩展功能。

2．本章介绍的《贪吃蛇》游戏存在一个小问题：当玩家快速对蛇头变向时会发生错误导致失败。例如，当贪吃蛇向右移动时，快速按方向键↓、方向键←，或者方向键↑、方向键←，此时会判定为游戏失败。试解释出现该状况的原因，并提出解决方案。

3．（选做）本章介绍的《贪吃蛇》游戏的蛇头与蛇身的表达方式完全一样，请思考当蛇头与蛇身外观不一致时应该如何选择数据结构并实现，并分析该实现方式与原实现方式在时间效率上的差异。

第 12 章

五子棋

12.1 游戏简介

《五子棋》是一款两人对弈的策略游戏，不但简单易学、老少皆宜，而且趣味性十足。通过玩《五子棋》游戏，不仅可以锻炼玩家的思维能力，还可以提升玩家的智力和反应能力。《五子棋》游戏源于 4000 多年前，比围棋的历史还要悠久。在《五子棋》游戏中，玩家需要使用黑色和白色的棋子来竞争形成 5 个连成一条线的棋子，谁先成功连成 5 个棋子，谁便赢得了比赛。《五子棋》游戏的画面如图 12-1 所示。

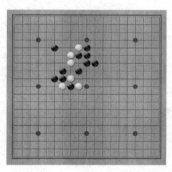

图 12-1 《五子棋》游戏的画面

12.2 游戏规则

12.2.1 《五子棋》游戏的棋盘和棋子

《五子棋》游戏的棋盘与围棋的棋盘相似，但比围棋的棋盘少 4 路，即由 15 路（共 225 个交叉点）组成。在棋盘上，上、下两端的横线称为端线，左、右两端的纵线称为边线。从两条端线和两条边线向正中发展而纵横交叉在第 4 条线形成的 4 个点称为"星"，棋盘正中一点称为"天元"。

以对局开始时的黑方为准，棋盘上的横线从近到远用阿拉伯数字 1~15 标记，纵线从左到右用英文字母 A~O 表示。由于每个英文字母都对应一条纵线，每个阿拉伯数字都对应一条横线，因此棋盘上的每个交叉点都可以用英文字母和阿拉伯数字的组合来表示。在标记一个交叉点时，一般把英文字母放在前面，阿拉伯数字放在后面，如"天元"的位置为 H8，4 个"星"的位置分别为 D4、L4、D12 和 L12，如图 12-2 所示。

《五子棋》游戏的棋子分为黑棋和白棋：黑棋 113 个，白棋 112 个，共计 225 个，恰好和棋盘上交叉点的数目相同。在对弈的过程中，双方的棋子可放置于空白的交叉点上，如图 12-3 所示。

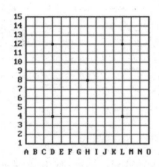

图 12-2　《五子棋》游戏的棋盘　　　　　　图 12-3　落子位置

12.2.2　《五子棋》游戏的基本规则

《五子棋》游戏的基本规则如下：黑白双方依次落子，先在棋盘上形成横向、纵向、斜向连续相同颜色的 5 个以上（含 5 个）棋子的一方为胜，如图 12-4 所示。

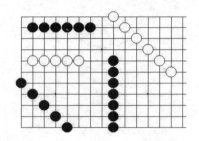

图 12-4　《五子棋》游戏的获胜条件

12.2.3　落子顺序

游戏开始时，落子顺序以黑方为先，白方为后。黑方的第 1 个棋子应下在"天元"（H8）位置，白方第 1 个棋子只能下在以"天元"为中心邻近的 8 个位置，黑方第 2 个棋子只能下在以"天元"为中心邻近的 5×5 个位置，白方第 2 个棋子不受任何限制，可下在棋盘上的任意位置。

12.3　游戏开发核心思路

12.3.1　绘制棋盘

绘制整个棋盘可以通过绘制一张棋盘的图片来完成。为了方便计算，每个交叉点相隔 1

个单位，即整个棋盘共有 15×15 个单位，如图 12-5 所示。

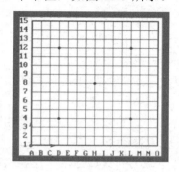

图 12-5　绘制棋盘

12.3.2　绘制棋子

使用一个 15×15 的二维数组来表示棋盘上棋子的位置。每个元素的数值分别表示无棋子、白棋和黑棋。例如，可以使用 0 表示无棋子，–1 表示白棋，1 表示黑棋。为了使程序具有可阅读性，也可以使用枚举类型来表示。

数组的下标表示交叉点的位置，以左下角为第 1 个交叉点，该位置对应的数组下标为 [0][0]，右上角对应的数组下标为 [14][14]。

这个二维数组是整个游戏的关键数据结构，不仅可以用于绘制当前整个棋盘的棋子位置，方便遍历该数组，还可以用于游戏的胜负判定。

12.3.3　落子

对局双方为交替落子，可以设置一个枚举类型来切换双方落子的权限。

通过获得鼠标指针在屏幕上的位置，并映射成世界坐标系中的位置，可以寻找到落子的位置，并计算出落子的位置所对应的数组元素序号。

为了方便计算，棋盘上的每个交叉点相隔 1 个单位，以棋盘左下角的第 1 个交叉点为原点。横向坐标为 X，纵向坐标为 Y，那么某个交叉点的二维数组的坐标为 $[X][Y]$。假设当前鼠标指针在棋盘上的坐标为 (M_X, M_Y)，因为鼠标指针的具体位置不是正整数，所以利用四舍五入（Round 方法）计算鼠标指针坐标的整数值，并映射到盘面棋形的二维数组的坐标中，即 $[X][Y] = [Round(M_X), Round(M_Y)]$，这样玩家就不需要非常准确地单击落子点。例如，鼠标指针当前在 H8 位置附近，如 (7.3,7.2)，那么映射到盘面棋形的二维数组的坐标为 [7][7]（注意：数组的起始坐标为 [0][0]）。

在每次落子后，通过修改表示盘面棋形的二维数组，可以保存当前的盘面棋形；通过遍历整个二维数组来判定胜负。

对于重新开始的游戏，只需要对盘面棋形的二维数组清零即可。

12.3.4　获胜规则

《五子棋》游戏是通过确定横向、纵向、反斜向和正斜向是否有 5 个相同颜色的棋子相连

来判定胜负的。下面以"—"、"|"、"\"和"/"分别表示横向、纵向、反斜向和正斜向。当一方落子结束后将进行连五判定。

- 搜索盘面棋形二维数组中"—"上的数据是否有连续相同颜色的 5 个棋子，如果有，则说明有一方构成了连五，否则转到下一条。
- 搜索盘面棋形二维数组中"|"上的数据是否有连续相同颜色的 5 个棋子，如果有，则说明有一方构成了连五，否则转到下一条。
- 搜索盘面棋形二维数组中"\"上的数据是否有连续相同颜色的 5 个棋子，如果有，则说明有一方构成了连五，否则转到下一条。
- 搜索盘面棋形二维数组中"/"上的数据是否有连续相同颜色的 5 个棋子，如果有，则说明有一方构成了连五，否则转到下一条。

如果以上几个方向均没有连续相同颜色的 5 个棋子，则说明当前盘面棋形二维数组中没有连五的情况，转为禁手判断，如果没有禁手，则双方可以继续落子。

12.3.5　判断黑方禁手功能

当黑方落子后，判断黑方是否出现三三禁手、四四禁手和长连禁手等情况，可以按照以下逻辑进行。

- 获得当前棋子落下的位置，并修改盘面棋形二维数组。
- 搜索盘面棋形二维数组中与落子位置相连的"—"上的数据是否构成禁手，如果已构成，则返回，否则转到下一条。
- 搜索盘面棋形二维数组中与落子位置相连的"|"上的数据是否构成禁手，如果已构成，则返回，否则转到下一条。
- 搜索盘面棋形二维数组中与落子位置相连的"\"上的数据是否构成禁手，如果已构成，则返回，否则转到下一条。
- 搜索盘面棋形二维数组中与落子位置相连的"/"上的数据是否构成禁手，如果已构成，则返回，否则转到下一条。
- 如果都没有构成，则转为探索与落子位置不相连的，即搜索中间有一个空格的数据。
- 搜索盘面棋形二维数组中落子位置有一个空格的"—"上的数据是否构成禁手，如果已构成，则返回，否则转到下一条。
- 搜索盘面棋形二维数组中落子位置有一个空格的"|"上的数据是否构成禁手，如果已构成，则返回，否则转到下一条。
- 搜索盘面棋形二维数组中落子位置有一个空格的"\"上的数据是否构成禁手，如果已构成，则返回，否则转到下一条。
- 搜索盘面棋形二维数组中落子位置有一个空格的"/"上的数据是否构成禁手，如果已构成，则返回，否则转向下一条。
- 如果都没有构成，则说明没有禁手，由白方继续落子。

12.3.6　游戏流程图

游戏流程图如图 12-6 所示。

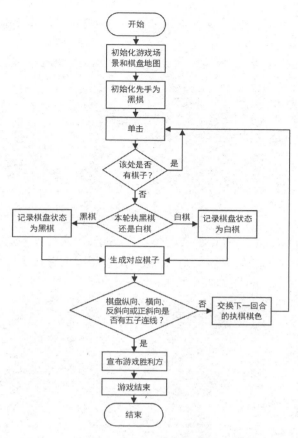

图 12-6　游戏流程图

12.4　游戏实现

12.4.1　前期准备

（1）新建工程。新建一个名为 Gomoku 的工程，选择 2D 选项，如图 12-7 所示。

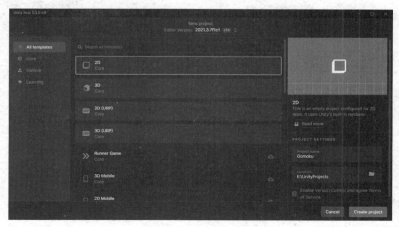

图 12-7　新建工程

（2）导入图片资源。创建不同的文件夹用来保存不同的资源。将下载的图片保存到 Sprites 文件夹中，如图 12-8 所示。

图 12-8 导入图片资源

（3）调整图片的 Pixels Per Unit 参数的值。在 Sprites 文件夹中选择 ChessBoard 图片、Black_Chess 图片和 White_Chess 图片，在 Inspector 窗口中调整这 3 张图片的 Pixels Per Unit 参数的值，如图 12-9 所示，将图片大小与 Unity 中的单位对应。

图 12-9 调整图片的 Pixels Per Unit 参数的值

（4）设置棋盘的位置和摄像机的视野范围。在 Sprites 文件夹中选择 ChessBoard 图片，将其拖入场景中，并在 Inspector 窗口中调整它的 Transform 组件的 Position 参数，如图 12-10 所示。调整摄像机的视野范围，使背景图片占据整个视野。在 Hierarchy 窗口中单击 Main Camera 物体后，在 Inspector 窗口中调整 Size 参数的值，图 12-11 所示。

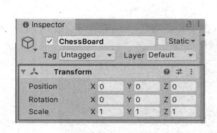

图 12-10 设置棋盘的位置

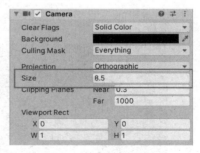

图 12-11 设置摄像机的视野范围

12.4.2 初始化棋盘地图

（1）设置棋盘的范围。新建 4 个空物体，并将其分别命名为 TL、TR、BL 和 BR。将这 4 个空物体分别调整到棋盘右上角、左上角、右下角和左下角的位置，如图 12-12 所示。

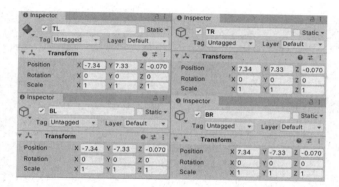

图 12-12 设置棋盘的范围

（2）在 Scripts 文件夹中新建 GameController 脚本，其中用一个 15×15 的二维数组来存储棋盘上所有可以落子的位置。已知棋盘左上角的坐标为（x,y），棋盘上每格的宽度为 width，高度为 height，在数组中索引号为[row,col]的位置的坐标可以根据如下公式计算：

$$\begin{cases} \text{pos}x = \text{width} \times (\text{col} - 7) \\ \text{pos}y = \text{height} \times (\text{row} - 7) \end{cases}$$

GameController 脚本中的代码如下：

```
using System.Collections;
using System.Collections.Generic;
using UnityEngine;
using UnityEngine.UI;

public class GameController : MonoBehaviour
{
    public Transform top_left;              //左上角空物体，用于标记位置
    public Transform top_right;             //右上角空物体，用于标记位置
    public Transform bottom_left;           //左下角空物体，用于标记位置
    public Transform bottom_right;          //右下角空物体，用于标记位置
    Vector2 pos_TL,pos_TR,pos_BL,pos_BR;    //左上角、右上角、左下角和右下角的坐标值
    float gridWith,gridHeight;              //棋盘上一格的宽度和高度
    Vector2 [,] chessPos;                   //可以放置棋子的位置
    public enum ChessInfo{Blank = 0, Black = 1, White = -1}
    //棋盘上每格的状态，分别为空、黑棋和白棋
    ChessInfo chessInfo;
    ChessInfo[,] snapshotMap;    //棋盘状态

    //Start is called before the first frame update
    void Start()
    {
        InitChessBoard();
    }

    //Update is called once per frame
    void Update()
    {
        PutChess();
    }

    void InitChessBoard(){
```

```csharp
        chessPos = new Vector2[15,15];              //分配棋子位置二维数组内存空间
        snapshotMap = new ChessInfo[15,15];         //分配棋盘快照内存空间
        //获得左上角、右上角、左下角和右下角位置的数据
        pos_TL = top_left.position;
        pos_TR = top_right.position;
        pos_BL = bottom_left.position;
        pos_BR = bottom_right.position;
        //计算每格的宽度和高度
        gridWith = (pos_TR.x - pos_TL.x) / 14;
        gridHeight = (pos_TR.y - pos_BR.y) / 14;

        //计算每个落子点的位置,并保存到数组中
        for(int row =0; row < 15; row++){
            for(int col = 0; col < 15; col++){
                //把每个落子点的位置保存到二维数组中
                chessPos[row,col] = new Vector2(gridWith * (col - 7),
                                               gridHeight * (row-7));
                //测试数据
                GameObject obj = new GameObject(row+" " + col);
                //注意y轴方向上的负号,因为数组索引号与y轴的方向刚好相反
                obj.transform.position = new Vector3(chessPos[row,col].x,
                                                    -chessPos[row,col].y,0);
            }//end col
        }//end row
    }
}
```

（3）挂载脚本并赋值。新建一个空物体并将其命名为 ChessBoard，为该物体挂载 GameController 脚本。将之前创建的 BR、BL、TR 和 TL 拖到 ChessBoard 物体下，使其成为其子物体，并将这 4 个物体赋给 GameController 脚本，如图 12-13 所示。

图 12-13　挂载脚本并赋值

12.4.3　编写落子框架

对 GameController 脚本进行修改。通过 Camera.main.ScreenToWorldPoint()函数获得鼠标单

击在场景中的位置，并且利用曼哈顿距离求出离鼠标指针最近的落子点。GameController 脚本中的代码如下：

```csharp
using UnityEngine.UI;

public class GameController : MonoBehaviour
{
    ......
    public float threshold = 0.2f;
    Vector2 mousePos;

    //Update is called once per frame
    void Update()
    {
        PutChess();
    }
    ......
    void PutChess(){
        if(Input.GetMouseButton(0)){
            //把鼠标指针在屏幕上的坐标位置转换成世界坐标
            mousePos = Camera.main.ScreenToWorldPoint(Input.mousePosition);
            //Debug.DrawRay(mousePos,Camera.main.transform.forward,Color.red);
            //Debug.Log(Input.mousePosition + " " + mousePos);
            //找到鼠标指针的位置与棋盘上的哪个落子点最近
            for(int row = 0; row < 15;row++){
                for(int col = 0; col < 15; col++){
                    //找到最近的落子点
                    if(IsNearBy(mousePos, chessPos[row,col])){
                        //Debug.Log("mousePos = " + mousePos + " ChessPoint = " +
                                chessPos[row,col]);
                        //在该位置创建棋子
                        //CreateChess(row,col,chessPos[row,col]);
                        //判断是否有五子相连
                        //DetectionWiner();
                    }
                }//end col
            }//end row
        }
    }

    bool IsNearBy(Vector2 mousePos, Vector2 chessPoint){
        //计算鼠标指针位置与落子点之间的曼哈顿距离，若该距离值小于阈值则表示两者相近，返回true
        if(Mathf.Abs(chessPoint.x-mousePos.x)+Mathf.Abs(chessPoint.y-mousePos.y)
            < threshold)
        {
            return true;
        }else{
            return false;
        }
    }
}
```

12.4.4 落子

（1）创建棋子，且只有在空棋位才可以创建，并切换落子权限。现在需要做的是黑白双方交替落子。可以用枚举类型表示黑白双方。在黑方落子后，切换为白方落子状态；在白方落子后，切换为黑方落子状态。在 GameController 脚本中修改对应的代码：

```
using UnityEngine.UI;
public class GameController : MonoBehaviour
{
    ……

    public enum Turn{Black,White}
    Turn turn = Turn.Black;                        //默认黑棋为先
    public GameObject blackChess;                   //黑棋对象
    public GameObject whiteChess;                   //白棋对象
    ……

    //Update is called once per frame
    void Update()
    {
        PutChess();
    }
    ……

    void PutChess(){
        if(Input.GetMouseButton(0)){
            //把鼠标指针的屏幕坐标转换成世界坐标
            mousePos = Camera.main.ScreenToWorldPoint(Input.mousePosition);
            //Debug.DrawRay(mousePos,Camera.main.transform.forward,Color.red);
            //Debug.Log(Input.mousePosition + " " + mousePos);
            //找到鼠标指针的位置与棋盘上哪个落子点的距离最近
            for(int row = 0; row < 15;row++){
                for(int col = 0; col < 15; col++){
                    //找到最近的落子点
                    if(IsNearBy(mousePos, chessPos[row,col])){
                        //Debug.Log("mousePos = " + mousePos + " ChessPoint = " +
                            chessPos[row,col]);
                        //在该位置创建棋子
                        CreateChess(row,col,chessPos[row,col]);
                    }
                }//end col
            }//end row
        }
    }

    void CreateChess(int row, int col, Vector2 putPosition){
        //如果该处没有棋子
        if(snapshotMap[row,col] == ChessInfo.Blank){
            //修改快照，把该空位置设置为对应落子方的状态
            snapshotMap[row,col]=(turn==Turn.Black?ChessInfo.Black:
                                ChessInfo.White);
            //根据落子方生成对应的棋子
            switch(turn){
                case Turn.Black:
                    //生成黑棋
```

```
                Instantiate(blackChess,new Vector3(putPosition.x,
                    putPosition.y,0),Quaternion.identity).name
                    = "Black " + row.ToString() + "_" + col.ToString();
                //交换落子方
                turn = Turn.White;
            break;
            case Turn.White:
                //生成白棋
                Instantiate(whiteChess,new Vector3(putPosition.x,
                    putPosition.y,0),Quaternion.identity).name
                    = "White " + row.ToString() +"_" + col.ToString();
                //交换落子方
                turn = Turn.Black;
            break;
        }
    }
}
```

（2）修改 Order in Layer 参数的值，并为 GameController 脚本赋值。将黑棋和白棋制作成预制体，并将两个预制体的 Order in Layer 参数设置为 1，同时为 GameController 脚本赋值，如图 12-14 和图 12-15 所示。

图 12-14　修改 Order in Layer 参数的值

图 12-15　为 GameController 脚本赋值

（3）测试黑白双方交替落子的功能。运行游戏，这时已经可以实现黑白双方交替落子，如图 12-16 所示。

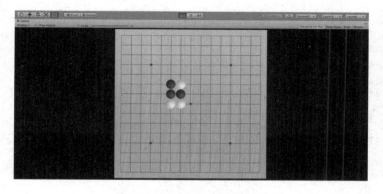

图 12-16　黑白双方交替落子

12.4.5　连子判断和胜负判断

（1）黑白双方是否连五的判断。连五一共有 4 种情况，分别为横连、纵连、正斜连和反斜连。在 GameController 脚本中修改对应的代码：

```csharp
using UnityEngine.UI;

public class GameController : MonoBehaviour
{
    ......
    void CreateChess(int row, int col, Vector2 putPosition){
        //如果该处没有棋子
        if(snapshotMap[row,col] == ChessInfo.Blank){
            //修改快照，把该空位置设置为对应落子方的状态
            snapshotMap[row,col] = (turn == Turn.Black ? ChessInfo.Black : ChessInfo.White);
            //根据落子方生成对应的棋子
            switch(turn){
                case Turn.Black:
                    //生成黑棋
                    Instantiate(blackChess,new Vector3(putPosition.x,putPosition.y,0),
                                Quaternion.identity).name= "Black " + row.ToString() +
                                "_" + col.ToString();
                    //判断是否为黑方胜利
                    DetectionWiner();
                    //交换落子方
                    turn = Turn.White;
                    break;
                case Turn.White:
                    //生成白棋
                    Instantiate(whiteChess,new Vector3(putPosition.x,putPosition.y,0),
                                Quaternion.identity).name ="White " + row.ToString() +
                                "_" + col.ToString();
                    //判断是否为白方胜利
                    DetectionWiner();
                    //交换落子方
                    turn = Turn.Black;
                    break;
            }
        }
    }

    void DetectionWiner(){
        switch(turn){
            case Turn.Black:
                //横向、纵向、正斜向和反斜向的探测，下面黑方和白方的代码类似
                //是否可以重构函数
                If(horizontalDetect(ChessInfo.Black) || VerticalDetect(ChessInfo.Black) ||
                    SlashDetect(ChessInfo.Black) || BackSlashDetect(ChessInfo.Black)){
                        Debug.Log("Black Win");
                }
                break;
            case Turn.White:
                //横向、纵向、正斜向和反斜向的探测
                if(horizontalDetect(ChessInfo.White) || VerticalDetect(ChessInfo.White) ||
                    SlashDetect(ChessInfo.White) || BackSlashDetect(ChessInfo.White)){
                        Debug.Log("White Win");
                }
```

```
        break;
    }//end switch
}

//横向探测
bool horizontalDetect(ChessInfo chessInfo){
    for(int row = 0; row < 15; row++){
        for(int col = 0; col < 11;col++){     //注意索引号
            //从当前棋子开始向右 4 个落子点，思考以下代码是否有可以优化的方法
            If(snapshotMap[row,col] == chessInfo &&
                snapshotMap[row,col+1] == chessInfo &&
                snapshotMap[row,col+2] == chessInfo &&
                snapshotMap[row,col+3] == chessInfo &&
                snapshotMap[row,col+4] == chessInfo)
            {
                return true;
            }
        }//end col
    }//end row
    return false;
}
//纵向探测
bool VerticalDetect(ChessInfo chessInfo){
    for(int row = 0; row < 11; row++){
        for(int col = 0; col < 15; col++){
            if(snapshotMap[row,col] == chessInfo &&
                snapshotMap[row+1,col] == chessInfo &&
                snapshotMap[row+2,col] == chessInfo &&
                snapshotMap[row+3,col] == chessInfo &&
                snapshotMap[row+4,col] == chessInfo)
            {
                return true;
            }
        }//end col
    }//end row
    return false;
}
//正斜向探测
bool SlashDetect(ChessInfo chessInfo){
    for(int row = 0; row < 11; row++){                      //注意索引值
        for(int col = 4; col < 15; col++){                  //注意索引值
            if(snapshotMap[row,col] == chessInfo &&
                snapshotMap[row+1,col-1] == chessInfo &&
                snapshotMap[row+2,col-2] == chessInfo &&
                snapshotMap[row+3,col-3] == chessInfo &&
                snapshotMap[row+4,col-4] == chessInfo)
            {
                return true;
            }
        }//end col
    }//end row
    return false;
}
```

```csharp
//反斜向探测
bool BackSlashDetect(ChessInfo chessInfo){
    for(int row = 0; row < 11;row++){              //注意索引值
        for(int col = 0; col < 11;col++){          //注意索引值
            if(snapshotMap[row,col] == chessInfo &&
                snapshotMap[row+1,col+1] == chessInfo &&
                snapshotMap[row+2,col+2] == chessInfo &&
                snapshotMap[row+3,col+3] == chessInfo &&
                snapshotMap[row+4,col+4] == chessInfo){
                return true;
            }
        }
    }
    return false;
}
```

（2）测试连五效果。运行游戏，当相同颜色的棋子形成连五后，在 Console 窗口中输出获胜方，如图 12-17 所示。

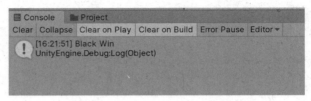

图 12-17　测试连五效果

（3）为了可以更直观地看到获胜方，这里用 UI 来提示获胜方。在 GameController 脚本中修改对应的代码：

```csharp
using UnityEngine.UI;

public class GameController : MonoBehaviour
{
    ……

    public Image winImage;                  //用于挂载胜利画面的 ImageUI
    public Sprite blackWinSprite;           //黑方胜利画面
    public Sprite whiteWinSprite;           //白方胜利画面
    private bool isGameOver = false;

    ……
    void Update()
    {
        if(isGameOver==false)
        {
            PutChess();
        }
    }

    void DetectionWiner(){
        switch(turn){
```

```csharp
    case Turn.Black:
    //横向、纵向、正斜向和反斜向的探测,下方黑方和白方的代码类似,是否可以重构函数?
    if(horizontalDetect(ChessInfo.Black) || VerticalDetect(ChessInfo.Black) ||
      SlashDetect(ChessInfo.Black) || BackSlashDetect(ChessInfo.Black))
        {
        isGameOver = true;
        winImage.gameObject.SetActive(true);
        winImage.sprite = blackWinSprite;
        Debug.Log("Black Win");
    }
    break;
    case Turn.White:
    //横向、纵向、正斜向和反斜向的探测
    if(horizontalDetect(ChessInfo.White) || VerticalDetect(ChessInfo.White) ||
      SlashDetect(ChessInfo.White) || BackSlashDetect(ChessInfo.White))
        {
        isGameOver = true;
        winImage.gameObject.SetActive(true);
        winImage.sprite = whiteWinSprite;
        Debug.Log("White Win");
    }
    break;
    }//end switch
  }
}
```

（4）把图片资源赋给 GameController 脚本。新建 Image 对象,将其赋给 GameController 脚本,将一张胜利图片赋给该 Image 对象,如图 12-18 所示,并调整图片大小;隐藏该 Image 对象;把两张胜利图片赋给 GameController 脚本,图 12-19 所示。

（5）测试游戏效果。运行游戏,最终效果如图 12-20 所示。

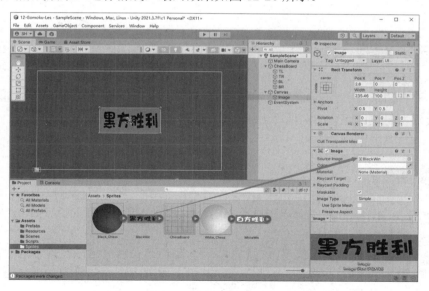

图 12-18　新建 Image 对象

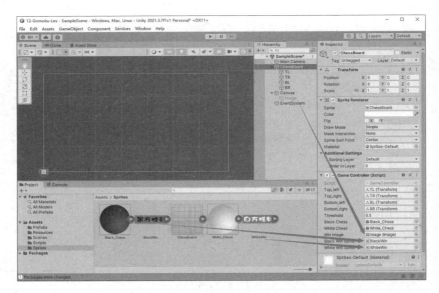

图 12-19　把两张胜利图片赋给 GameController 脚本

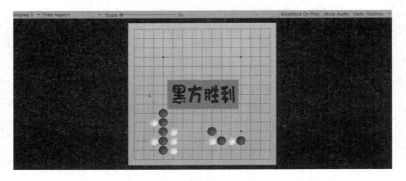

图 12-20　最终效果

12.5　本章小结

本章主要介绍了《五子棋》游戏的设计与实现，其中重点在于实现规则的判定，由于实现规则相对较为复杂，因此在实现时需要格外细心。

12.6　练习题

1. 本章介绍的连五的判定方法为每次遍历整个棋盘以寻找在横向、纵向、正斜向和反斜向是否存在相同颜色 5 个相连的棋子，该方法实现简单，试分析其时间复杂度。

2. 在上一道题目的基础上思考改进算法。因为每次落子时只可能在落子点附近出现相同颜色 5 个相连棋子的情况，所以只需要在当前落子点周围 8 个方向检测一小段距离即可。试比较该算法与上一道题目算法的异同，并尝试实现该算法。

3. 试实现简单的长连禁手，即判断黑方是否出现 6 个或 6 个以上棋子相连（可采用上一道题目的思路）。

第 13 章

跳棋

13.1 游戏简介

《跳棋》游戏，又叫《波子棋》游戏，是正方跳棋的改良版。1892 年，德国的 Ravensburger 公司便取得了这款游戏的专利，并命名为 *Sternhalma*，意为"星形跳棋"。这是一款可以由 2～6 个玩家同时进行的棋类游戏，棋盘为六角星形，棋子有 6 种颜色。每个玩家占据一个角，使用同一种颜色的棋子。《跳棋》是一项广受欢迎的益智型棋类游戏，无论是老人还是小孩都可以玩。《跳棋》游戏的画面如图 13-1 所示。

图 13-1 《跳棋》游戏的画面

13.2 游戏规则

《跳棋》游戏的规则（"跳"的示意图）如图 13-2 所示，棋子在有直线连接的相邻 6 个方向一步一步移动，如果相邻位置上有任意一方的一个棋子，且该位置直线方向的下一个位置是空的，则可以直接"跳"到该空位上。在"跳"的过程中，只要相同条件满足便可以连续进行。谁最先把正对面的"阵地"全部占领，谁便取得胜利。

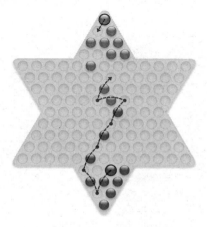

图 13-2 《跳棋》游戏的规则

13.3 游戏开发核心思路

13.3.1 棋盘排列

《跳棋》游戏的棋盘较为特殊，为六角星形，使用常用的数据结构难以表达此类布局，因此需要跳出常规思维，从其他角度思考问题，尽可能把不规则的布局转化为可用常规数据结构表达的形式。通过观察发现，可以将棋盘坐标设置为如图 13-3 所示的形式。

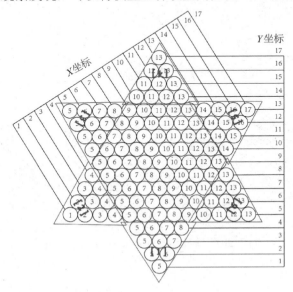

图 13-3 棋盘坐标形式

这样做的好处是：从 X 坐标映射到 Y 坐标，Y 坐标值是连续的。当 X 坐标值为 1 时，Y 坐标值只能为 5；当 X 坐标值为 2 时，Y 坐标值的集合为{5,6}；当 X 坐标值为 3 时，Y 坐标值的集合为{5,6,7}。以此类推，以便建立二维数组。若按照普通的笛卡儿坐标系，则会出现不连续的 Y 坐标值，不方便编写程序。按照此方式建立二维数组的代码如下：

```
private int[,] cellsMap = {
    {5,5},              //x 为 1, y 的上限是 5, 下限是 5
    {5,6},              //x 为 2, y 的上限是 5, 下限是 6
    {5,7},              //x 为 3, y 的上限是 5, 下限是 7
    {5,8},              //x 为 4, y 的上限是 5, 下限是 8
    {1,13},             //x 为 5, y 的上限是 1, 下限是 13
    {2,13},             //x 为 6, y 的上限是 2, 下限是 13
    {3,13},             //x 为 7, y 的上限是 3, 下限是 13
    {4,13},             //x 为 8, y 的上限是 4, 下限是 13
    {5,13},             //x 为 9, y 的上限是 5, 下限是 13
    {5,14},             //x 为 10, y 的上限是 5, 下限是 14
    {5,15},             //x 为 11, y 的上限是 5, 下限是 15
    {5,16},             //x 为 12, y 的上限是 5, 下限是 16
    {5,17},             //x 为 13, y 的上限是 5, 下限是 17
    {10,13},            //x 为 14, y 的上限是 10, 下限是 13
    {11,13},            //x 为 15, y 的上限是 11, 下限是 13
    {12,13},            //x 为 16, y 的上限是 12, 下限是 13
```

```
            {13,13},            //x为17, y的上限是13, 下限是13
};
```

根据上面的数组,可以很容易地写出一个方法,根据 X 和 Y 的值来确定该位置是否合法(是否在棋盘内)。如果获取的 X 和 Y 的值不在区间内,则返回 false;否则返回 true。pos$[X-1,0]$为 Y 的上限,pos$[X-1,0]$为 Y 的下限。判断该位置是否合法的代码如下:

```
bool IsLegalPosition(int x,int y){
    if(x < 1 || x > 17){
        return false;
    }
    if(y < cellsMap[x-1,0] || y > cellsMap[x-1,1]){
        return false;
    }
    return true;
}
```

13.3.2 生成棋子

在构建好棋盘的表达形式后,可以根据棋盘位置坐标生成棋子。如果在图 13-3 中位置[1]处生成棋子,则对应的 X 坐标的取值范围为 5~8,对应的 Y 坐标的取值范围为 1~4,可以编写如下代码框架生成对应棋子:

```
for(int i=x 坐标下限; i<=x 坐标上限; i++)
{
    for(int j=y 坐标下限; j<=y 坐标上限; i++)
    {
        if(位置存在)
        {
            //创建新棋子
        }
    }
}
```

13.3.3 棋子在 Unity 中笛卡儿坐标系下的位置映射

将棋子和棋盘位置都设定各自的 X 坐标值和 Y 坐标值(不同于空间坐标中的 x 和 y,这里的 X 和 Y 代表棋盘位置坐标)。

设某个棋格在自定义坐标系下的坐标为 P_{custom}(x_{custom}, y_{custom}),因为 Unity 中使用的是标准的笛卡儿坐标系,所以设该棋格在笛卡儿坐标系中的位置为 P_{pos}(x_{pos}, y_{pos})。在 Unity 中生成棋格时,需要把 P_{custom} 转换为 P_{pos},可以通过以下转换关系构造出函数:

$$P_{pos} = f(P_{custom}) => \begin{cases} x_{pos} = a_1 \times x_{custom} + b_1 \times y_{custom} + c_1 \\ y_{pos} = a_2 \times x_{custom} + b_2 \times y_{custom} + c_2 \end{cases}$$

这里以中心位置[9,9]为原点,即它的空间坐标为(0,0),如图 13-4 所示。选取与中心位置[9,9]相邻的两个位置([8,9]和[8,8]),用 X 坐标值和 Y 坐标值分别对应空间中的 x 值和 y 值,联立三元一次方程组,求出 a_1、b_1 和 c_1,以及 a_2、b_2 和 c_2,由此可以求得棋盘位置坐标与空间坐标的映射关系。方程组如下:

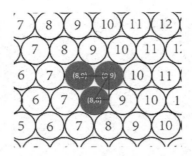

图 13-4 求解棋格在自定义坐标系下到 Unity 笛卡儿坐标系下的映射关系

对于中心位置[9,9]，有

$$\begin{cases} 9a_1 + 9b_1 + c_1 = 0 \\ 9a_2 + 9b_2 + c_2 = 0 \end{cases}$$

对于位置[8,9]，有

$$\begin{cases} 8a_1 + 9b_1 + c_1 = -2 \\ 8a_2 + 9b_2 + c_2 = 0 \end{cases}$$

对于位置[8,8]，有

$$\begin{cases} 8a_1 + 8b_1 + c_1 = -1 \\ 8a_2 + 8b_2 + c_2 = -\sqrt{3} \end{cases}$$

由以上公式（当然也可以尝试使用线性代数的相关知识求解）可以求得

$$\begin{cases} a_1 = 2, b_1 = -1, c_1 = -9 \\ a_2 = 0, b_1 = \sqrt{3}, c_1 = -9\sqrt{3} \end{cases}$$

求解后得到，若物体的位置为[X,Y]，则物体的空间位置坐标 pos 可以表示为

$$\text{pos} = (2 \times X - Y - 9, \sqrt{3} \times (Y - 9))$$

13.3.4 计算可移动位置

当有棋子被选中时，选中的棋子周围各位置的坐标关系如图 13-5 所示。

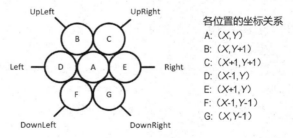

图 13-5 棋子周围各位置的坐标关系

先为每个棋盘位置设置一个布尔类型的变量 canMove，初始值为 false，只有在可移动时值才为 true。

要寻找棋子可连续跳动的位置需要向棋子的 6 个方向寻找，逐个方向寻找可跳动位置。在寻找每个方向的跳动位置时，要实现棋子的隔子跳动，也意味着要找到一个棋子，并且以这个棋子做跳板，跳到同方向对称的位置上。因此，在寻找某个方向是否有可跳动位置时可分为 3 步判断，此处设某个方向为 i。

第 1 步：判断 i 方向是否有棋子。如果没有棋子，则该方向位置可走，将状态设为可移动，因为没有执行过跳动，所以不用继续判断；如果有棋子，则该棋子便是跳板，进入第 2 步判断。

第 2 步：判断当前棋子 i 方向的棋子的 i 方向位置是否有棋子。如果有棋子，则退出（说明位置被占住，当前棋子不能跳到此位置）；如果没有棋子，则该位置可跳动，所以将该位置设为可移动状态，此时进入第 3 步判断。

第 3 步：判断第 2 步中的棋子除去跳过来的方向，剩余 5 个方向是否有棋子。如果 j 方向没有棋子，则第 2 步中位置为此方向最终跳动位置；如果有棋子，说明又有跳板，则跳到第 2 步，判定这个棋子 j 方向有没有棋子，进入继续循环判定。

需要注意的是，在程序中并没有直接获取棋子跳过来的方向，处理方法是进行 6 个方向的判定：在第 2 步中加一个判断条件，如果没有棋子且该位置没有被设为可移动状态，则进入第 3 步。如果该位置已经被设为可移动状态，则说明已经判定过那个位置，现在进行此次判定是从第 3 步"反跳"回来的。进行 6 个方向的判定，如果不加这个判断条件，则会出现死循环，始终在第 2 步和第 3 步之间跳转。

13.3.5 回合限制

以有 2 个玩家为例，可以设置一个变量 round，从 1 开始计数，每回合结束加 1。如果变量 round 的值除以 2，余数为 1，则是玩家 1 的回合；如果余数为 0，则是玩家 2 的回合。如果有 4 个玩家，则除以 4，判断余数，以此类推（本章以有 2 个玩家展开介绍）。

13.3.6 游戏胜负判断

如果将一方棋子全部移动到对面则取得胜利。因此在每步落子结束后，需要进行胜负判断。每个棋子都有与位置相对应的 X 坐标值和 Y 坐标值。可以编写一个方法，如果一方开始生成在 X 坐标值为 5～8，Y 坐标值为 1～4 的位置，则目标区域是 X 坐标值为 10～13，Y 坐标值为 14～17 的位置。每下完一步，便判断各个目标区域是否存在棋子，并且都是以它为目标一方的棋子。若都成立，则这一方取得胜利。

13.3.7 游戏流程图

游戏流程图如图 13-6 所示。

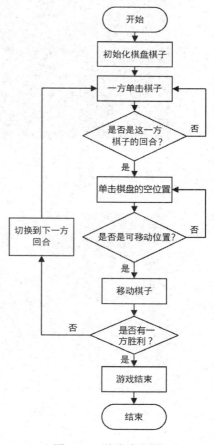

图 13-6　游戏流程图

13.4 游戏实现

13.4.1 前期准备

(1) 新建工程。新建一个名为 ChineseCheckers 的 2D 工程，如图 13-7 所示。

(2) 导入图片资源。创建不同的文件夹用来保存不同的资源，将下载的图片资源保存到 Sprites 文件夹中，如图 13-8 所示。

(3) 修改摄像机参数。选中摄像机，设置 Inspector 窗口中的参数，如图 13-9 所示。

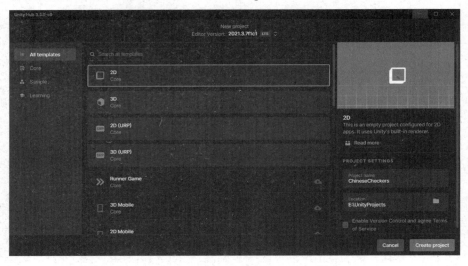

图 13-7　新建工程

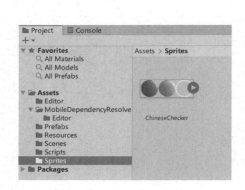

图 13-8　导入图片资源

图 13-9　修改摄像机参数

(4) 修改图片参数。选中 ChineseChecker 图片，设置 Inspector 窗口中的参数，如图 13-10 所示。

(5) 切割棋子图片。选中 ChineseChecker 图片，在 Inspector 窗口中，先单击 Sprite Editor 按钮，再单击 Slice 下拉按钮，将 Type 设置为 Grid By Cell Count（见图 13-11），并且将其拆分成 3 个部分，分别命名为 RedChess、GreenChess 和 Cell，如图 13-12 所示。

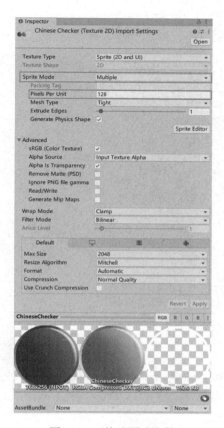

图 13-10 修改图片参数

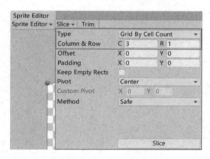

图 13-11 设置 Type 参数

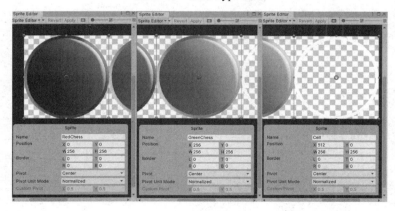

图 13-12 棋子图片命名

13.4.2 创建棋格

(1) 在 Scripts 文件夹下,新建一个脚本并将其命名为 ChessBoard,此脚本用来控制整个游戏流程。下面实现初始化棋盘功能。先打开 ChessBoard 脚本,再编辑该脚本。ChessBoard 脚本中的代码如下:

```
public class ChessBoard : MonoBehaviour
{
    public GameObject cellPrefab;
    private Cell[,] map = new Cell[18,18];
    private int[,] cellsMap = {
        {5,5},        //x 为 1, y 的上限是 5, 下限是 5
        {5,6},        //x 为 2, y 的上限是 5, 下限是 6
        {5,7},        //x 为 3, y 的上限是 5, 下限是 7
        {5,8},        //x 为 4, y 的上限是 5, 下限是 8
        {1,13},       //x 为 5, y 的上限是 1, 下限是 13
        {2,13},       //x 为 6, y 的上限是 2, 下限是 13
        {3,13},       //x 为 7, y 的上限是 3, 下限是 13
        {4,13},       //x 为 8, y 的上限是 4, 下限是 13
        {5,13},       //x 为 9, y 的上限是 5, 下限是 13
        {5,14},       //x 为 10, y 的上限是 5, 下限是 14
        {5,15},       //x 为 11, y 的上限是 5, 下限是 15
        {5,16},       //x 为 12, y 的上限是 5, 下限是 16
        {5,17},       //x 为 13, y 的上限是 5, 下限是 17
        {10,13},      //x 为 14, y 的上限是 10, 下限是 13
        {11,13},      //x 为 15, y 的上限是 11, 下限是 13
        {12,13},      //x 为 16, y 的上限是 12, 下限是 13
        {13,13},      //x 为 17, y 的上限是 13, 下限是 13
    };

    //Start is called before the first frame update
    void Start()
    {
        //1.创建棋盘
        CreateCellMap();
        //2.创建棋子

    }

    //Update is called once per frame
    void Update()
    {

    }

    void CreateCellMap(){
        for(int x = 0; x <= 17; x++){
            for(int y = 0; y <= 17; y++){
                if(IsLegalPosition(x,y)){
                    GameObject cell = Instantiate(cellPrefab);
                    cell.name = "Cell:" + x + " " + y;
                    map[x,y] = cell.GetComponent<Cell>();
                    map[x,y].SetCellPos(x,y);
                    SetCartesianPosition(cell,x,y,0);
                }
```

```
                }//end y
            }//end x
        }

        bool IsLegalPosition(int x,int y){
            if(x < 1 || x > 17){
                return false;
            }
            if(y < cellsMap[x-1,0] || y > cellsMap[x-1,1]){
                return false;
            }
            return true;
        }

        void SetCartesianPosition(GameObject cell, int x, int y,int z){
            cell.transform.position = new Vector3(2*x-y-9,Mathf.Sqrt(3)*(y-9),z);
        }
}
```

（2）创建 GameController 物体。先将 Sprites 文件夹中的 Cell 制作成预制体，再在场景中创建一个空物体并将其命名为 GameController，最后为这个物体添加 ChessBoard 脚本。在 Inspector 窗口中找到 ChessBoard 脚本，并拖入 Prefabs 文件夹下的 Cell，如图 13-13 所示。

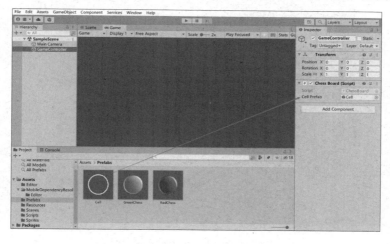

图 13-13　拖入 Prefabs 文件夹下的 Cell

（3）单击"运行"按钮，可以看到自动生成了棋盘，如图 13-14 所示。

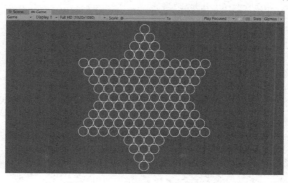

图 13-14　自动生成的棋盘

13.4.3 创建棋子

（1）在 Scripts 文件夹下，新建一个脚本并将其命名为 Chess，该脚本用来保存棋子位置信息。先打开 Chess 脚本，再编辑该脚本。Chess 脚本中的代码如下：

```
public class Chess : MonoBehaviour
{
    public int customPos_x, customPos_y;

    public void SetCustomPosition(int x, int y){
        customPos_x = x;
        customPos_y = y;
    }
}
```

（2）为棋子预制体添加脚本。将 Sprites 文件夹中的 GreenChess 和 RedChess 制作成预制体，为 Prefabs 文件夹下的 GreenChess 和 RedChess 添加 Chess 脚本，如图 13-15 所示。

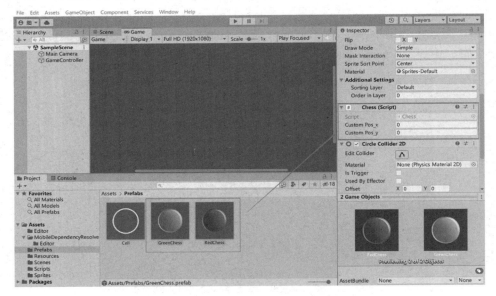

图 13-15　为棋子预制体添加脚本

（3）在 ChessBoard 脚本中添加生成棋子的代码。ChessBoard 脚本中的代码如下（已经说明的代码用省略号代替，新添加的为加粗字体）：

```
public class ChessBoard : MonoBehaviour
{
    //……
    public GameObject redChess, greenChess;
    //……
    void Start()
    {
        //1.创建棋盘
        CreateCellMap();
        //2.创建棋子
        CreateChesses(redChess, 5, 8, 1, 4);
```

```
        CreateChesses(greenChess, 10, 13, 14, 17);
    }

    //……

    void CreateChesses(GameObject chessPrefab, int x_upper, int x_lower,
                      int y_upper, int y_lower)
    {
        for (int x = x_upper; x <= x_lower; x++)
        {
            for (int y = y_upper; y <= y_lower; y++)
            {
                if (IsLegalPosition(x, y))
                {
                    GameObject chess = Instantiate(chessPrefab);
                    chess.GetComponent<Chess>().SetCustomPosition(x, y);
                    map[x, y].hasChess = true;
                    SetCartesianPosition(chess, x, y, -1);
                    map[x, y].chessTag = chess.tag;
                }
            }//end y
        }//end x
    }
```

（4）将预制体拖到对应处。在保存代码后，选中 GameController 物体，在 Inspector 窗口中找到 ChessBoard 脚本，并拖入 Prefabs 文件夹下的 GreenChess 和 RedChess，如图 13-16 所示。

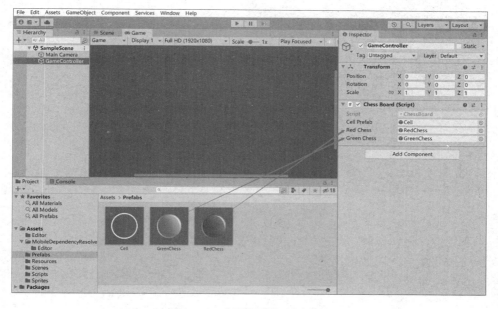

图 13-16　将预制体拖到对应处

（5）单击"运行"按钮，在棋盘上自动生成的棋子如图 13-17 所示。

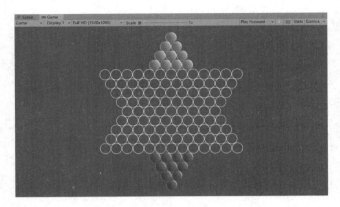

图 13-17　在棋盘上自动生成的棋子

13.4.4　选择棋子

（1）为棋子添加高亮显示。先为 Prefabs 文件夹中的 GreenChess 预制体和 RedChess 预制体分别添加 Sprites 文件夹中 Chinese Checker 图片组的 Cell 图片为子物体，并设置参数，然后关闭 Cell 的 Sprite Renderer 脚本，将其高亮框隐藏，如图 13-18 所示。

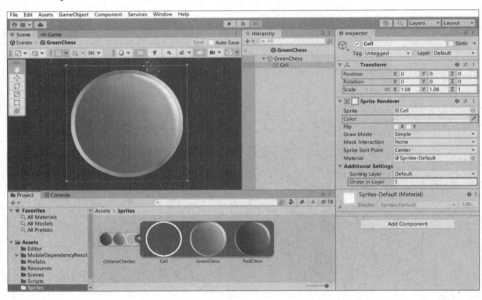

图 13-18　为棋子添加高亮显示

（2）利用射线进行检测。通过标签判断单击的物体是什么，如果是棋子，则高亮显示。ChessBoard 脚本中的代码如下：

```
public class ChessBoard : MonoBehaviour
{
    //……
    private GameObject selectedChess;
    //……
    void Update()
    {
        SelectChess();
```

```
}
//……
void SelectChess()
{
    if (Input.GetMouseButtonDown(0))
    {
        RaycastHit2D hit = Physics2D.Raycast(Camera.main.
                                    ScreenToWorldPoint(Input.mousePosition),
                                        Camera.main.transform.forward);
        if (hit.collider != null)
        {
            if (hit.transform.tag == "RedChess" || hit.transform.tag ==
                "GreenChess")
            {
                ChangeChessHighlight(selectedChess);
                selectedChess = hit.transform.gameObject;
                ChangeChessHighlight(selectedChess);
            }
        }
    }
}

void ChangeChessHighlight(GameObject selectedChess)
{
    if (selectedChess == null)
    {
        return;
    }
    SpriteRenderer highlightRing = selectedChess.transform.GetChild(0)
                                        .GetComponent<SpriteRenderer>();
    highlightRing.enabled = !highlightRing.enabled;
}
```

（3）添加碰撞盒和标签。先为 Prefabs 文件夹下的 GreenChess、RedChess 和 Cell 分别添加 Circle Collider 2D 组件，再为 GreenChess、RedChess 和 Cell 分别赋予标签 GreenChess、RedChess 和 Cell，如图 13-19 所示。

图 13-19　添加碰撞盒和标签

（4）单击"运行"按钮，若单击棋子，则棋子会高亮显示。棋子高亮显示的效果如图 13-20 所示。

彩色图

图 13-20 棋子高亮显示的效果

13.4.5 判断邻近可走棋格

（1）在 Scripts 文件夹下，新建一个脚本并将其命名为 Cell，该脚本用来保存棋盘位置信息和设置棋盘位置状态。Cell 脚本中的代码如下：

```
public class Cell : MonoBehaviour
{
    public int customPos_X, customPos_Y;
    public bool canMove = false;
    public bool hasChess = false;
    public Color normalColor = Color.white;
    public Color highlightColor = Color.yellow;
    private SpriteRenderer sr;
    //Start is called before the first frame update
    void Start()
    {
        sr = GetComponent<SpriteRenderer>();
    }

    //Update is called once per frame
    void Update()
    {
        ChangeHighLightColor();
    }

    public void SetCellPos(int x, int y)
    {
        customPos_X = x;
        customPos_Y = y;
    }

    public void ChangeHighLightColor()
    {
        if (canMove == true)
        {
            sr.color = highlightColor;
        }
        else
```

```
            {
                sr.color = normalColor;
            }
        }
    }
}
```

(2) 判断棋子周围邻近棋格是否可走。如果可走，则高亮显示可走棋格。返回 Scripts 文件夹，打开 ChessBoard 脚本进行修改，ChessBoard 脚本中的代码如下：

```
public class ChessBoard : MonoBehaviour
{
    //……
    public enum Direction { UpLeft, UpRight, Left, Right, DownLeft, DownRight }
    struct NextPos { public int x; public int y; }
    private NextPos nextPos;

    //……

    void SelectChess()
    {
        if (Input.GetMouseButtonDown(0))
        {
            RaycastHit2D hit = Physics2D.Raycast(Camera.main.ScreenToWorldPoint(
                                        Input.mousePosition),
                                        Camera.main.transform.forward);
            if (hit.collider != null)
            {
                if (hit.transform.tag == "RedChess" || hit.transform.tag == "GreenChess")
                {
                    ChangeChessHighlight(selectedChess);
                    selectedChess = hit.transform.gameObject;
                    ChangeChessHighlight(selectedChess);
                    Chess chessComp = selectedChess.GetComponent<Chess>();
                    DetectAllowedMovePlace(chessComp.customPos_x,
                                        chessComp.customPos_y);

                }
            }
        }
    }

    //……

    void DetectAllowedMovePlace(int curPosX, int curPosY)
    {
        ClearCanMoveTag();
        //遍历邻近 6 个方向是否可走
        AdjacentCellDetection(curPosX, curPosY, Direction.UpLeft);
        AdjacentCellDetection(curPosX, curPosY, Direction.UpRight);
        AdjacentCellDetection(curPosX, curPosY, Direction.Left);
        AdjacentCellDetection(curPosX, curPosY, Direction.Right);
        AdjacentCellDetection(curPosX, curPosY, Direction.DownLeft);
        AdjacentCellDetection(curPosX, curPosY, Direction.DownRight);
    }
```

```csharp
    void AdjacentCellDetection(int curPosX, int curPosY, Direction dir)
    {
        NextPos nextPos = GetNextPos(curPosX, curPosY, dir);
        if (!IsLegalPosition(nextPos.x, nextPos.y))
        {
            return;
        }
        Cell selectedCell;
        selectedCell = map[nextPos.x, nextPos.y];
        if (selectedCell.hasChess == false && selectedCell.canMove == false &&
map[curPosX, curPosY].hasChess == true)
        {
            selectedCell.canMove = true;
            return;
        }
    }

    NextPos GetNextPos(int curPosX, int curPosY, Direction dir)
    {
        switch (dir)
        {
            case Direction.UpLeft:
                nextPos.x = curPosX;
                nextPos.y = curPosY + 1;
                break;
            case Direction.UpRight:
                nextPos.x = curPosX + 1;
                nextPos.y = curPosY + 1;
                break;
            case Direction.Left:
                nextPos.x = curPosX - 1;
                nextPos.y = curPosY;
                break;
            case Direction.Right:
                nextPos.x = curPosX + 1;
                nextPos.y = curPosY;
                break;
            case Direction.DownLeft:
                nextPos.x = curPosX - 1;
                nextPos.y = curPosY - 1;
                break;
            case Direction.DownRight:
                nextPos.x = curPosX;
                nextPos.y = curPosY - 1;
                break;
        }
        return nextPos;
    }

    void ClearCanMoveTag()
    {
        for (int x = 0; x <= 17; x++)
```

```
            {
                for (int y = 0; y <= 17; y++)
                {
                    if (map[x, y] != null)
                    {
                        map[x, y].canMove = false;
                    }
                }
            }
        }//ClearCanMoveTag
}
```

（3）测试棋子可走棋格的高亮显示。将 Cell 脚本挂载到 Cell 预制体下，单击"运行"按钮，单击棋子，可以看到棋子可走的棋格高亮显示，如图 13-21 所示。

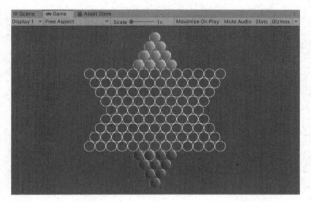

图 13-21 棋子可走的棋格高亮显示

彩色图

13.4.6 单步移动棋子

（1）在 SelectChess() 函数中多加一个判断，用于在选择棋子后选择空棋格进行移动。MoveChess() 函数用于移动棋子和整理棋格的状态。ChessBoard 脚本中的代码如下：

```
public class ChessBoard : MonoBehaviour
{
    //……
    void SelectChess()
    {
        if (Input.GetMouseButtonDown(0))
        {
            RaycastHit2D hit = Physics2D.Raycast(Camera.main.ScreenToWorldPoint(
                                            Input.mousePosition),
                                            Camera.main.transform.forward);
            if (hit.collider != null)
            {
                if (hit.transform.tag == "RedChess" || hit.transform.tag == "GreenChess")
                {
                    ChangeChessHighlight(selectedChess);
                    selectedChess = hit.transform.gameObject;
                    ChangeChessHighlight(selectedChess);
                    Chess chessComp = selectedChess.GetComponent<Chess>();
```

```
                    DetectAllowedMovePlace(chessComp.customPos_x,
                                           chessComp.customPos_y);
            }
            else if (hit.transform.tag == "Cell" && selectedChess != null)
            {
                Cell selectedCell = hit.transform.GetComponent<Cell>();
                if (selectedCell.canMove == true)
                {
                    MoveChess(selectedCell);
                    ClearCanMoveTag();
                }
            }
        }
    }
}
void MoveChess(Cell selectedCell)
{
    //1.移动棋子
    int movePosX = selectedCell.customPos_X;
    int movePosY = selectedCell.customPos_Y;
    selectedCell.hasChess = true;
    SetCartesianPosition(selectedChess, movePosX, movePosY, -1);
    ChangeChessHighlight(selectedChess);
    //2.对原来棋子位置的棋格状态进行清理
    Chess chessInfo = selectedChess.GetComponent<Chess>();
    map[chessInfo.customPos_x, chessInfo.customPos_y].hasChess = false;
    chessInfo.SetCustomPosition(movePosX, movePosY);
    selectedChess = null;
}
```

（2）测试单步移动棋子的效果。单击"运行"按钮，移动棋子，可以看到棋子可走的棋格会高亮显示，如图 13-22 所示。

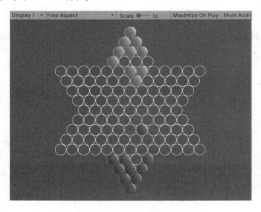

图 13-22 单步移动棋子的效果

彩色图

13.4.7 跳子

（1）使用递归思想循环判断棋子是否可以隔子跳动。ChessBoard 脚本中的代码如下：

```
public class ChessBoard : MonoBehaviour
```

```
{
    //......
    void AdjacentCellDetection(int curPosX, int curPosY, Direction dir)
    {
        NextPos nextPos = GetNextPos(curPosX, curPosY, dir);
        if (!IsLegalPosition(nextPos.x, nextPos.y))
        {
            return;
        }
        Cell selectedCell;
        selectedCell = map[nextPos.x, nextPos.y];
        if (selectedCell.hasChess == false && selectedCell.canMove == false)
        {
            selectedCell.canMove = true;
            return;
        }
        else if (selectedCell.hasChess == true)
        {
            JumpCheckerDetection(selectedCell.customPos_X,
                                 selectedCell.customPos_Y, dir);
        }
    }

    void JumpCheckerDetection(int chessPosX, int chessPosY, Direction dir)
    {
        NextPos tempNextPos = GetNextPos(chessPosX, chessPosY, dir);
        if (!IsLegalPosition(tempNextPos.x, tempNextPos.y))
        {
            return;
        }
        Cell cell = map[tempNextPos.x, tempNextPos.y];
        if (cell.hasChess == true)
        {
            return;
        }
        else
        {
            if (cell.canMove == false)
            {
                cell.canMove = true;
                //递归调用
                AdjacentCellDetection(cell.customPos_X, cell.customPos_Y, Direction.UpLeft);
                AdjacentCellDetection(cell.customPos_X, cell.customPos_Y, Direction.UpRight);
                AdjacentCellDetection(cell.customPos_X, cell.customPos_Y, Direction.Left);
                AdjacentCellDetection(cell.customPos_X, cell.customPos_Y, Direction.Right);
                AdjacentCellDetection(cell.customPos_X, cell.customPos_Y, Direction.DownLeft);
                AdjacentCellDetection(cell.customPos_X, cell.customPos_Y, Direction.DownRight);
            }
        }
    }
}
```

（2）棋子在隔子跳动后只能跳子而不能移动。按照上面的代码，会出现隔子跳动后还可以移动的问题。根据分析，只有在原本位置上有棋子才可以进行单步移动，所以需要添加一个判断。ChessBoard 脚本中的代码如下：

```csharp
public class ChessBoard : MonoBehaviour
{
    //……
    void AdjacentCellDetection(int curPosX, int curPosY, Direction dir)
    {
        NextPos nextPos = GetNextPos(curPosX, curPosY, dir);
        if (!IsLegalPosition(nextPos.x, nextPos.y))
        {
            return;
        }
        Cell selectedCell;
        selectedCell = map[nextPos.x, nextPos.y];
        if (selectedCell.hasChess == false && selectedCell.canMove == false &&
            map[curPosX, curPosY].hasChess == true)
        {
            selectedCell.canMove = true;
            return;
        }
        else if (selectedCell.hasChess == true)
        {
            JumpCheckerDetection(selectedCell.customPos_X, selectedCell.customPos_Y, dir);
        }
    }
}
```

（3）测试隔子跳动功能。单击"运行"按钮，移动棋子，可以看到棋子可走的棋格会高亮显示，如图 13-23 所示。

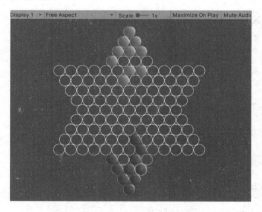

图 13-23　跳子功能的效果

彩色图

13.4.8　回合限制

（1）设置一个变量 round，每次行动后该变量的值加 1。可以利用取余（%）来切换红方和绿方回合。在 ChessBoard 脚本中添加两处代码：

```csharp
public class ChessBoard : MonoBehaviour
{
    //……
```

```csharp
private int round = 1;
//……
void MoveChess(Cell selectedCell)
{
    //1.移动棋子
    int movePosX = selectedCell.customPos_X;
    int movePosY = selectedCell.customPos_Y;
    selectedCell.hasChess = true;
    SetCartesianPosition(selectedChess, movePosX, movePosY, -1);
    ChangeChessHighlight(selectedChess);
    //2.对原来棋子位置的棋格状态进行清理
    Chess chessInfo = selectedChess.GetComponent<Chess>();
    map[chessInfo.customPos_x, chessInfo.customPos_y].hasChess = false;
    chessInfo.SetCustomPosition(movePosX, movePosY);
    selectedChess = null;
    round++;
}
```

（2）在 ChessBoard 脚本的 SelectChess()函数中,修改单击棋子后的判定条件。ChessBoard 脚本中的代码如下:

```csharp
void SelectChess()
{
    if (Input.GetMouseButtonDown(0))
    {
        RaycastHit2D hit = Physics2D.Raycast(Camera.main.ScreenToWorldPoint(
                                            Input.mousePosition),
                                            Camera.main.transform.forward);
        if (hit.collider != null)
        {
            if (hit.transform.tag == "RedChess" && round % 2 == 1 ||
                hit.transform.tag == "GreenChess" && round % 2 == 0)
            {
                ChangeChessHighlight(selectedChess);
                selectedChess = hit.transform.gameObject;
                ChangeChessHighlight(selectedChess);
                Chess chessComp = selectedChess.GetComponent<Chess>();
                DetectAllowedMovePlace(chessComp.customPos_x,
                                       chessComp.customPos_y);
            }
            else if (hit.transform.tag == "Cell" && selectedChess != null)
            {
                Cell selectedCell = hit.transform.GetComponent<Cell>();
                if (selectedCell.canMove == true)
                {
                    MoveChess(selectedCell);
                    ClearCanMoveTag();
                    WinDetection();
                }
            }
        }
    }
}
```

13.4.9 胜负判断

（1）在 ChessBoard 脚本中设置一个布尔类型的变量，用来表示游戏是否结束。确定游戏胜利的函数应该在 ChessBoard 脚本的 SelectChess()函数中调用，如果游戏结束就不能选择棋子。ChessBoard 脚本中的代码如下：

```
public class ChessBoard : MonoBehaviour
{
    //……
    private bool IsOver = false;
    //……
    void Update()
    {
        if (IsOver == false)
        {
            SelectChess();
        }
    }

    void SelectChess()
    {
        if (Input.GetMouseButtonDown(0))
        {
            RaycastHit2D hit = Physics2D.Raycast(Camera.main.ScreenToWorldPoint(
                                            Input.mousePosition),
                                            Camera.main.transform.forward);
            if (hit.collider != null)
            {
                if (hit.transform.tag == "RedChess" && round % 2 == 1 ||
                    hit.transform.tag == "GreenChess" && round % 2 == 0)
                {
                    ChangeChessHighlight(selectedChess);
                    selectedChess = hit.transform.gameObject;
                    ChangeChessHighlight(selectedChess);
                    Chess chessComp = selectedChess.GetComponent<Chess>();
                    DetectAllowedMovePlace(chessComp.customPos_x,
                                        chessComp.customPos_y);
                }
                else if (hit.transform.tag == "Cell" && selectedChess != null)
                {
                    Cell selectedCell = hit.transform.GetComponent<Cell>();
                    if (selectedCell.canMove == true)
                    {
                        MoveChess(selectedCell);
                        ClearCanMoveTag();
                        WinDetection();
                    }
                }
            }
        }
    }

    void WinDetection()
```

```csharp
{
    if (GameOver("RedChess", 10, 13, 14, 17))
    {
        IsOver = true;
        Debug.Log("红方胜利");
    }
    if (GameOver("GreenChess", 5, 8, 1, 4))
    {
        IsOver = true;
        Debug.Log("绿方胜利");
    }
}

bool GameOver(string tag, int x_upper, int x_lower, int y_upper, int y_lower)
{
    for (int x = x_upper; x <= x_lower; x++)
    {
        for (int y = y_upper; y <= y_lower; y++)
        {
            if (IsLegalPosition(x, y))
            {
                if (map[x, y].chessTag != tag)
                {
                    return false;
                }
            }
        }
    }
    return true;
}
}
```

（2）为 Cell 脚本增加变量，修改后的 Cell 脚本中的代码如下：

```csharp
public class Cell : MonoBehaviour
{
    public int customPos_X, customPos_Y;
    public bool canMove = false;
    public bool hasChess = false;
    public Color normalColor = Color.white;
    public Color highlightColor = Color.yellow;
    private SpriteRenderer sr;
    public string chessTag = "";
    //Start is called before the first frame update
    void Start()
    {
        sr = GetComponent<SpriteRenderer>();
    }

    //Update is called once per frame
    void Update()
    {
        ChangeHighLightColor();
    }
```

```
    public void SetCellPos(int x, int y)
    {
        customPos_X = x;
        customPos_Y = y;
    }

    public void ChangeHighLightColor()
    {
        if (canMove == true)
        {
            sr.color = highlightColor;
        }
        else
        {
            sr.color = normalColor;
        }
    }
}
```

（3）在 ChessBoard 脚本中增加代码，为棋子所在的棋格赋予标签。ChessBoard 脚本中的代码如下：

```
public class ChessBoard : MonoBehaviour
{
    void CreateChesses(GameObject chessPrefab, int x_upper, int x_lower,
                      int y_upper, int y_lower)
    {
        for (int x = x_upper; x <= x_lower; x++)
        {
            for (int y = y_upper; y <= y_lower; y++)
            {
                if (IsLegalPosition(x, y))
                {
                    GameObject chess = Instantiate(chessPrefab);
                    chess.GetComponent<Chess>().SetCustomPosition(x, y);
                    map[x, y].hasChess = true;
                    SetCartesianPosition(chess, x, y, -1);
                    map[x, y].chessTag = chess.tag;
                }
            }//end y
        }//end x
    }
    void MoveChess(Cell selectedCell)
    {
        //1.移动棋子
        int movePosX = selectedCell.customPos_X;
        int movePosY = selectedCell.customPos_Y;
        selectedCell.hasChess = true;
        SetCartesianPosition(selectedChess, movePosX, movePosY, -1);
        ChangeChessHighlight(selectedChess);
        **selectedCell.chessTag = selectedChess.tag;**
        //2.对原来棋子位置的棋格状态进行清理
        Chess chessInfo = selectedChess.GetComponent<Chess>();
        map[chessInfo.customPos_x, chessInfo.customPos_y].hasChess = false;
```